SAILING THE OPTIMIST

First published in Holland 1982 by Hollandia
First published in Great Britain 1987 by Nautical Books.
An imprint of Conway Maritime Press,
24 Bride Lane, Fleet Street,
London EC4Y 8DR

ISBN 0 85177 445 8

Cover design by Tony Garrett
Typeset by Witwell Ltd, Liverpool
Printed and bound in Great Britain by The Bath Press, Bath

SAILING THE OPTIMIST

MARJOLIJN & FEDDE SONNEMA,

THEO KEMPER, KAREL HEIJNEN

Contents

Introduction

This book is the result of the experience gained from instructing children of between ten and fourteen years to sail the Optimist dinghy. During the sailing courses we try to let novice sailors sail independently as much as possible.

When you start to sail, many questions arise and because of your enthusiasm you want to know all the answers at once. Now you can answer many of the questions with the help of this little book. It is not, therefore, only to be used on the sailing course but also at home.

You can also use this little book as a memory aid and read again at your leisure how a manoeuvre should be performed. Perhaps the book will give you new ideas which you can go and try on the water later.

We hope that it will help you get more pleasure from sailing.

Marjolijn Sonnema
Theo Kemper
Fedde Sonnema
Karel Heijnen

1
Parts of the Optimist

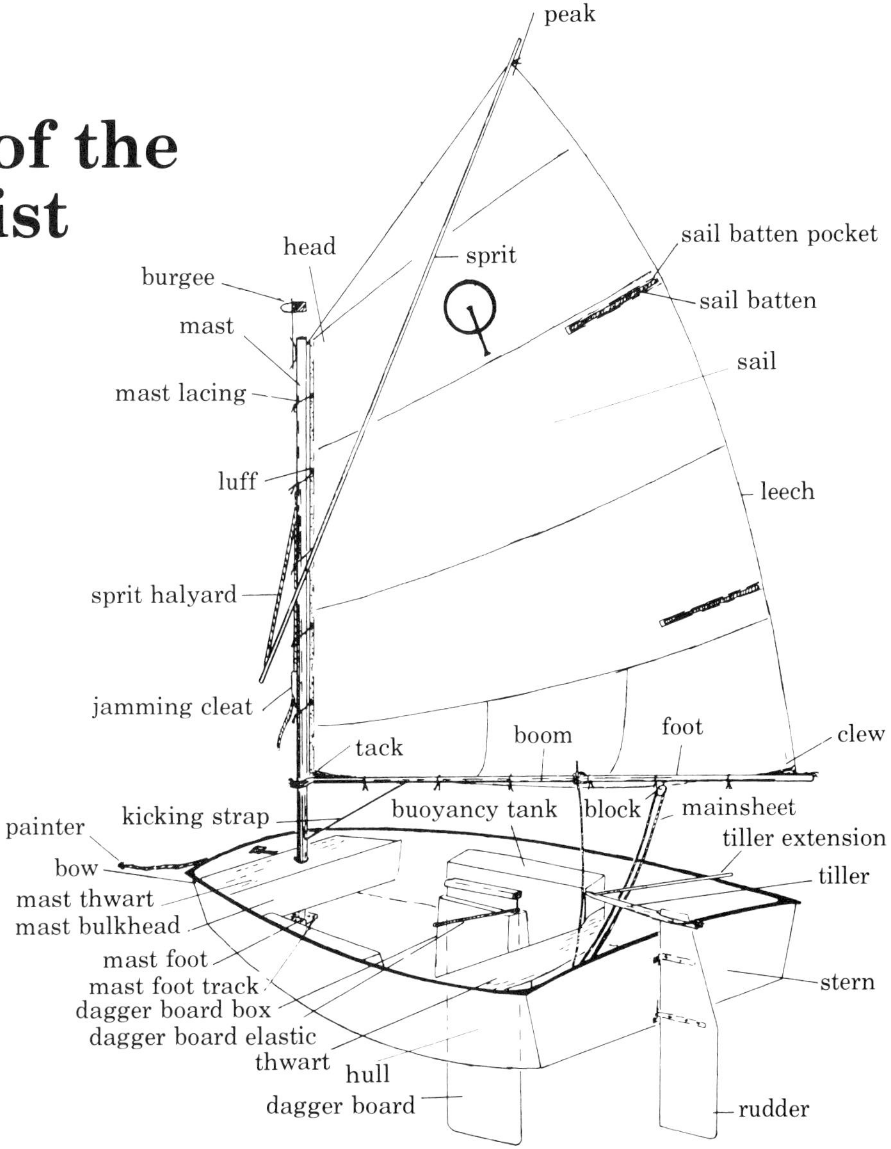

peak
head
sprit
sail batten pocket
burgee
sail batten
mast
sail
mast lacing
luff
leech
sprit halyard
jamming cleat
tack
boom
foot
clew
kicking strap
buoyancy tank
block
mainsheet
painter
tiller extension
bow
tiller
mast thwart
mast bulkhead
mast foot
mast foot track
stern
dagger board box
dagger board elastic
thwart
hull
dagger board
rudder

2
Look after yourself

Before reading about the boat and about sails, first something about yourself. What do you wear when sailing? There are certain clothes necessary in sailing. Just as much as in playing football or swimming. Clothes to keep you safe, warm and dry.

In the first place of course comes a lifejacket, even if you have a swimming certificate. There are a good many types of lifejacket. The best ones are the jackets that have a collar full of air or foam. This collar will keep your head above water. It must also have good fastenings.

Now, what do you put on under it? In most conditions you should wear a wind and waterproof sailing suit. You know that in an Optimist there is a lot of spray. Under this suit put on as many clothes as you think necessary to keep warm.

Now your feet. When choosing shoes you must remember two things:

1. The material must be supple, so that you can move easily in the boat.

2. The shoes should have non-slip soles. Gymshoes and supple lacing boots are most satisfactory for this. Bare feet are of course quite wrong: you can cut them terribly easily.

Finally a few more small points. If you have spectacles, put a piece of string on them so that you cannot lose them. Take off jewellery since it may get entangled. If you have long hair, put it up. It cannot then get stuck between sheet and sheaves or impede your view.

Remember
- **A good lifejacket.**
- **Water and windproof clothing.**
- **Warm clothing.**
- **Shoes with non-slip soles.**
- **String on spectacles.**
- **Jewellery off.**
- **Put up hair.**

3
Preparing the Optimist

On most boats you hoist the sail up the mast. With an Optimist you first attach the mast, the boom, the sprit and the sail to each other. After that you put all this (the rig) on the boat in one go. You only have to attach mast, boom, sprit and sail to each other the first time that you go sailing. After that you can keep all the rigging together and stow it neatly.

In order to make an Optimist ready to sail, you must therefore first prepare the rig and then set it up in the boat.

3.1 Preparing the rig

Before you start, it is best to look at a rigged Optimist. Then you can see how it all fits together.

Now you can begin. Lay the sail completely unfolded on a dry and clean place. Then lay the mast and the boom along the sail in their right places. The jaw of the boom must go round the mast. The sprit is laid over the mast and over the sail. You can see all this in the drawing. The numbers in the drawing indicate the sequence in which the corners must be attached. You can find the knots that you need for this in chapter 17, *Knots and hitches.* The cords with which you fasten the sail must be thin and quite supple, otherwise the knots may come undone. It is best to use 2mm twisted nylon. Furthermore, it is handy to fasten all these cords to the sail first with a reef knot. This prevents you from losing them all when you pull the sail from the mast, boom and sprit.

Next insert the sail battens.

Now the luff must be fastened to the mast and the foot to the boom. The outer ends are again fastened with reef knots. Make sure that the luff is drawn tight against the mast and the foot against the boom.

You have now fastened the sail everywhere that is necessary. There are a few things still to be done which you can also see in the drawing:

- Fasten the sprit halyard to the sprit with a bowline and pull it through the block and the jamming cleat.

- Fasten the kicking strap with a bowline to the eye on the boom and pull it through the jam cleat.
- Fasten the sheet with a clove hitch to the boom. At the outer end it is best to tie a half hitch as well, otherwise the clove hitch slips off.

You have now prepared the rig.

Corner 1

Put one cord round the mast and one round the boom. Fasten the outer end with a reef knot. Make sure that the corner of the sail is drawn tight against the mast and the boom.

Corner 2

Again use two cords, which are also fastened with reef knots. The eye in the sail must sit right next to the upper metal eye. The cord through the lower stirrup must not be drawn too tight.

Corner 3

You must fasten the eye in the sail to the metal eye on the sprit with a shackle.

Corner 4

You have two cords. One is fastened with a bowline to the eye in the sail. Pull that cord through the metal eye and after that again through the sail and again through the metal eye. Now pull the cord tight enough for the head to stand taut. Then fasten it with two half hitches. The other cord is fastened to the eye in the sail with a reef knot. Tie that round the boom and fasten it with a reef knot.

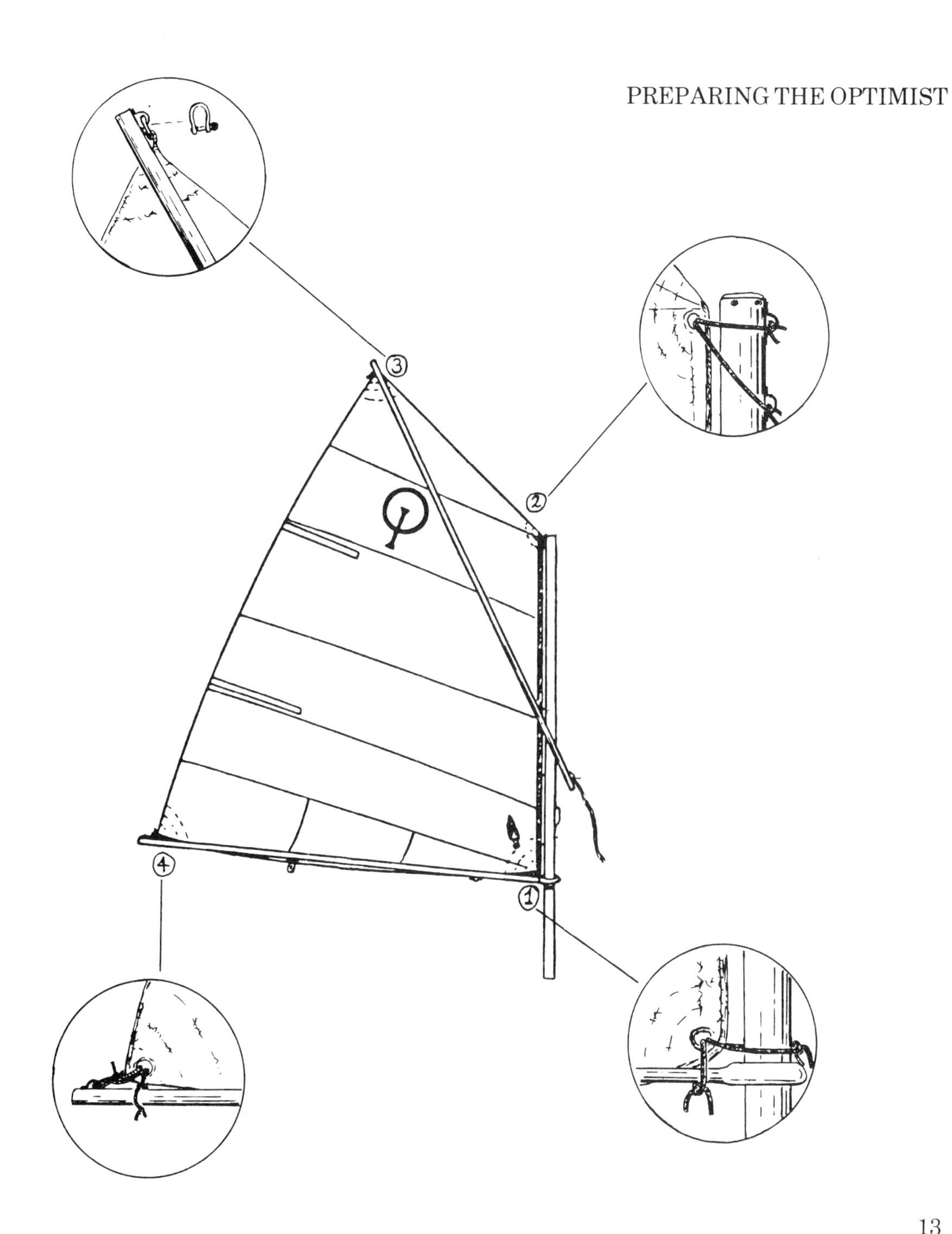

3
2
4
1

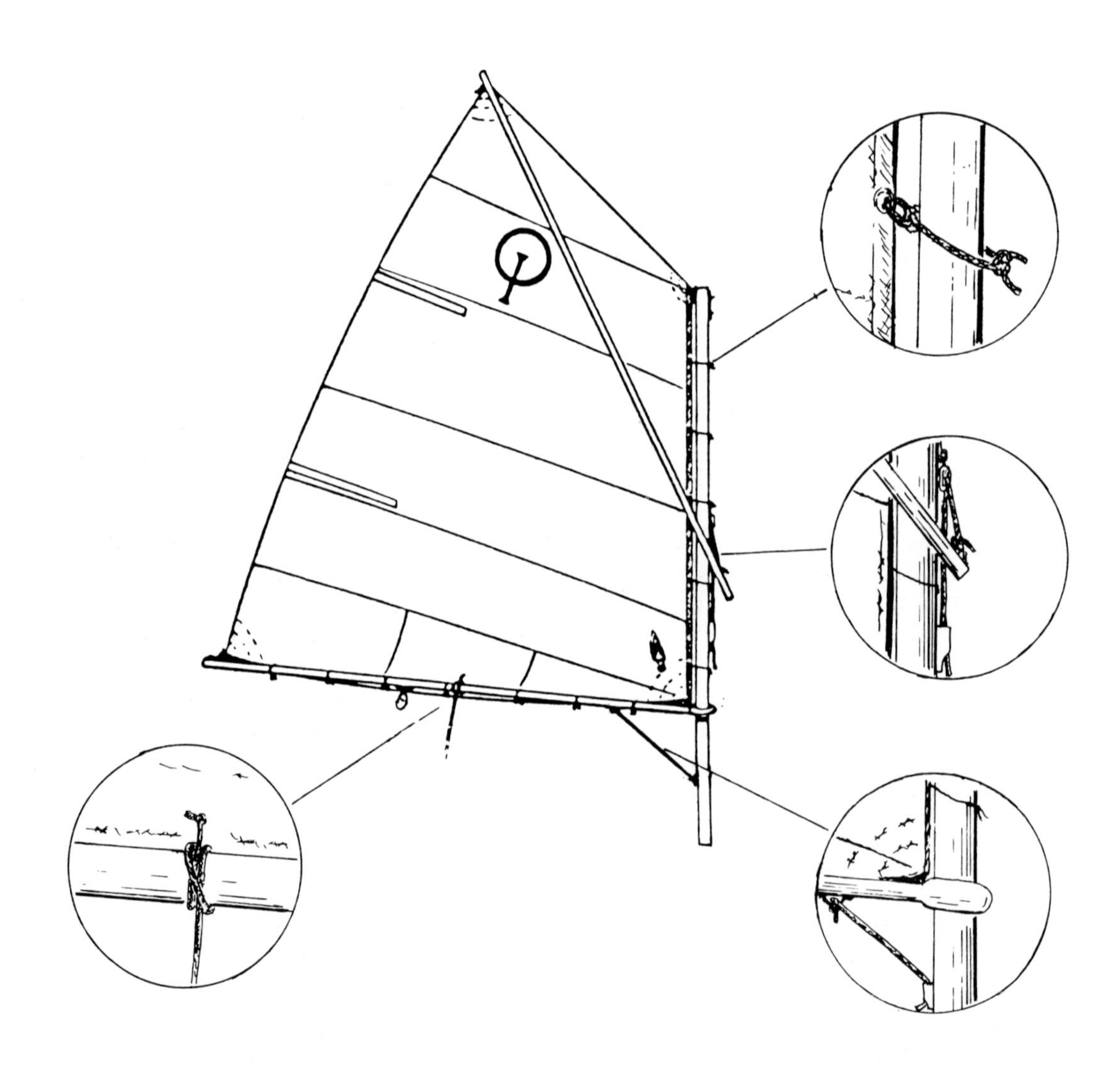

3.2 Rigging the boat

The best time to rig the boat is after it is launched but you should make sure that all the other gear is on board before slipping her into the water. So, first of all, put all the pieces of equipment that you will need into the boat. The dagger board and rudder are the most essential but do not forget the bailer and any other things like snacks or a can of lemonade.

You will need help to carry the boat into the water. At least one but better two friends will make the launching easier and prevent the boat getting bumped or scratched. Be careful to keep your back straight when lifting and bend your knees to take up the load.

Carry the boat to the water and launch it stern first, making sure that someone keeps hold of the painter which should be made fast once the boat is afloat.

Now you can hang the rudder in its place and prepare to put the rig into position.

If you have followed the previous steps the sail will be secured to the mast and you can therefore pick up the entire rig by yourself. Use both hands, spaced wide apart, to hold the mast in an upright position and remember to keep the sail angled into the wind. Now carry it to the boat and insert the mast through the hole in the mast thwart, making sure that the foot of the mast is secure in its track.

You can now climb on board and complete the rigging. First make sure that the mast foot is fixed so that it will not slip off the track. Now you can pull the sheet through the blocks as shown in the drawing. Do not forget to put a figure-of-eight knot at the end to prevent it flying out when you are sailing.

Finally, set the sail neatly by tightening the sprit halyard and the kicking strap. It is best to tighten the sprit halyard by pushing the sprit up with one hand while you are pulling the halyard through the jamming cleat with the other.

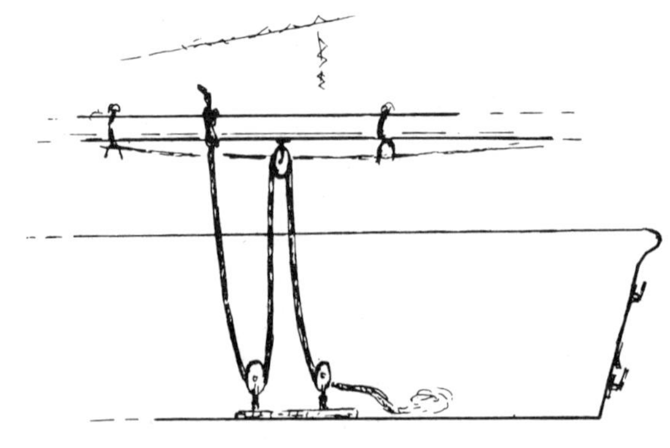
Reeving the sheet

In doing so you must tighten the halyard to the point where creases in the sail run neatly from the peak to the tack (see photo on page 17). After that, tighten the kicking strap by pressing the boom down with one hand and pulling the kicking strap through the jamming cleat with the other. The kicking strap must be tightened to the point where the creases from the tack to the peak partly disappear again.

Now insert the dagger board. Cast off and away you go!

Tighten the halyard to the point where creases in the sail run neatly from peak to tack. Then tighten the kicking strap.

4
Stowing the boat and rigging

As soon as you have come alongside and made fast, take the rig out of the boat. This prevents the sail flapping for a long time. Lay it down on a clean, dry place. Stowing this comes later.

Begin by stowing the boat. Take the rudder off and withdraw the dagger board. Then pull the boat onto the bank facing fore and aft. It is easier to lift the foremost section first as shown in the photograph.

Clean the boat with a sponge and fresh water and ensure that any sand or mud is removed. Now you can turn the boat over so that it does not fill with rain water.

Now tie the mast, boom and sprit together. You can see a clever way to stow the sail in the drawing.

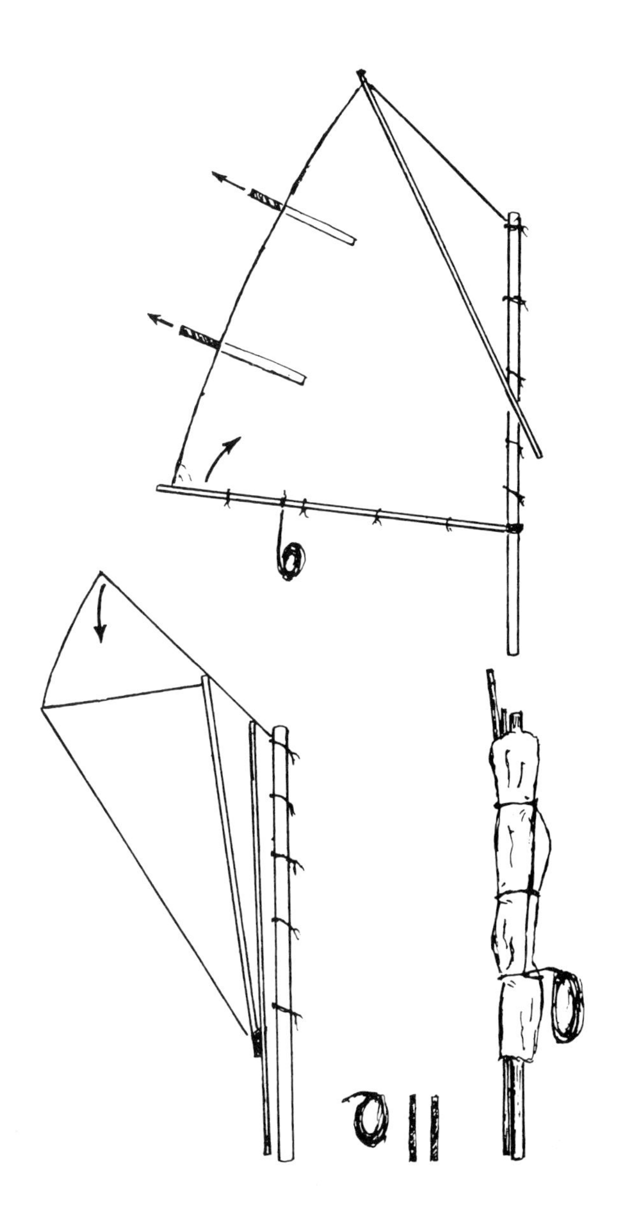

Pull the sail battens out of the sail, the kicking strap out of the jam cleat and the sheet away from the boom. Ease off the halyard a little, lay the boom sprit and the sprit alongside the mast and then fold the sail up.

Now roll the mast, boom and sprit up in the sail. Fasten the sheet to the end of the package with a clove hitch. Frap (see chapter 17 *Knots and hitches*) the sheet round the package and fasten it again to the other extremity of the whole with a clove hitch. (In chapter 22 *On a lee shore* you can read about how to stow the rigging in a faster manner.)

5 Steering

You steer a boat with a tiller. You can move to the left or right in a boat. This is the case only if the boat is moving. It can easily be compared with a bicycle. If you are standing still on a bicycle and you turn the handlebars to the left or right, you stay in the same place.

When steering yourself, you discover pretty quickly to which side the tiller must be turned in order to go to the left or right.

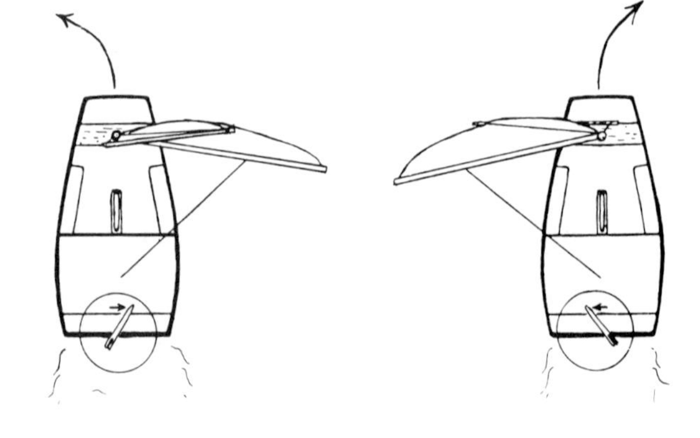

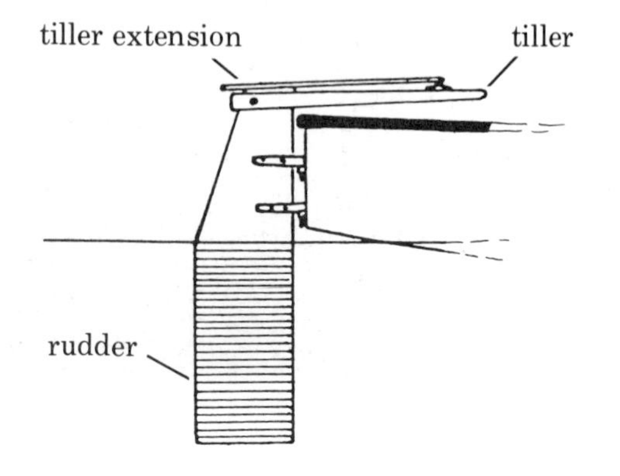

In order to be able to steer well, it is important to sit in the right position. You see in the drawing that the tiller projects into the boat. If you sit in the corner at the back of the boat, you cannot turn the tiller fully.

The best place to sit is with your hip against the wooden thwart, a little away from the tiller, so that you can easily move it to and fro.

But should you sit on the left or right

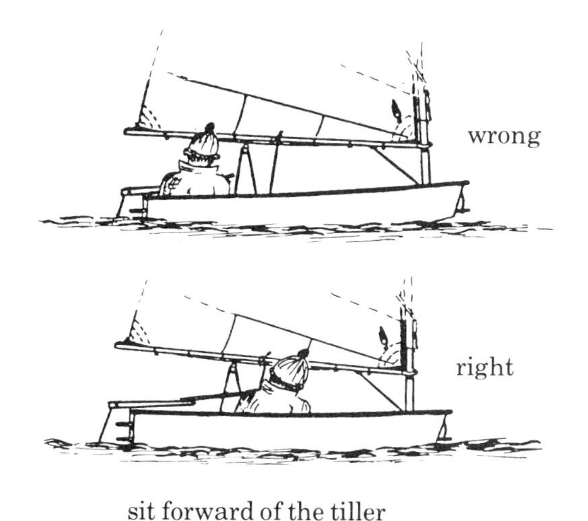

sit forward of the tiller

side of the boat? To decide you can use the following rule of thumb: Always sit on the side opposite the sail. Then the boat does not heel over so much. (Chapter 21.1 *Weight trimming*, also mentions that your position on board affects the speed of the boat.)

Incorrect helming position

Correct helming position

If you wish to sail in a straight line, take a fixed point on the land (a tree or rock) and keep it in line with the mast. If you wish to turn, find another reference point so that you have another object at which to aim.

A few more terms that are widely used: **windward** is the side *from which* the wind blows, **leeward** is the side *to which* the wind is blowing. If you look forwards in the boat, **port** is **left** and **starboard** is **right**.

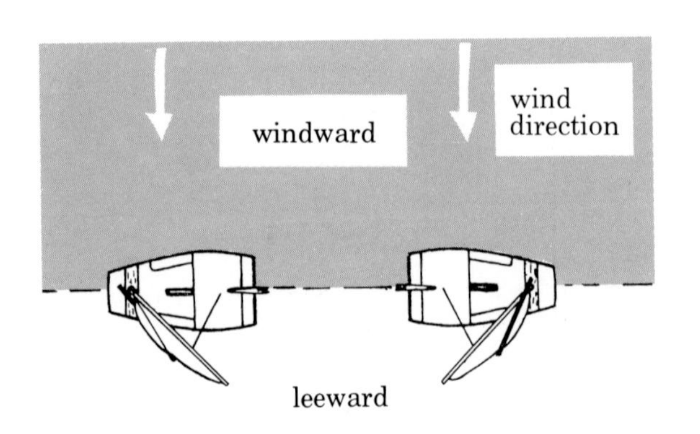

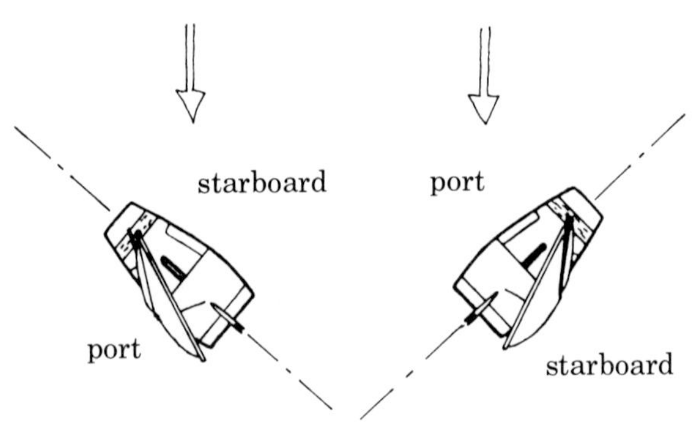

Remember

- **You must sit a little forward, so that the tiller can turn freely.**
- **Sit on the side opposite the sail.**

6
Points of sailing and use of the sheets

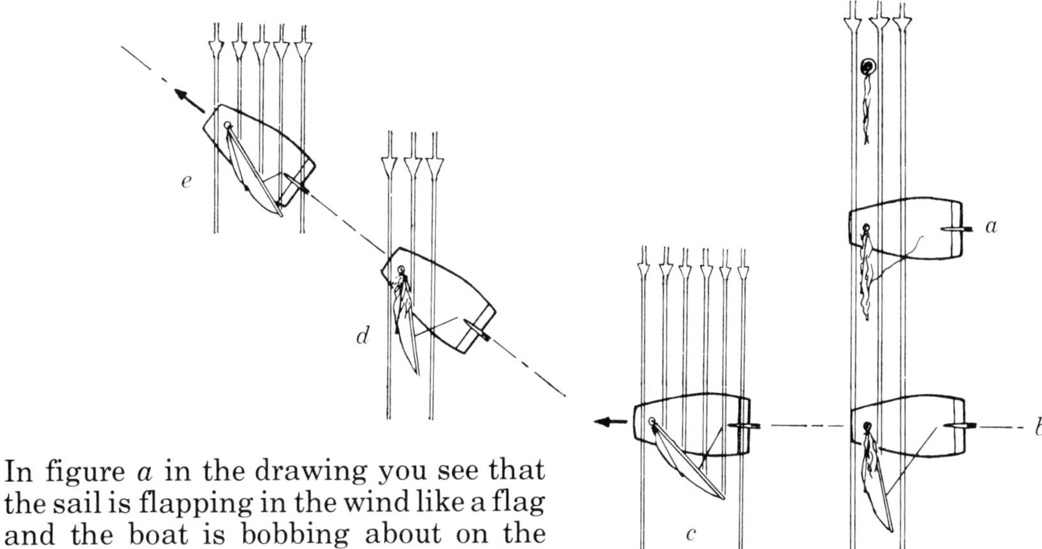

In figure *a* in the drawing you see that the sail is flapping in the wind like a flag and the boat is bobbing about on the water and not moving forward. *The direction of the wind is given in the drawing by the thin lines (wind arrows).* In order to move forward, you must pull in the sheet of the sail. If you do this slowly, you will see that the sail first catches wind at the outer edge and begins to billow out. The fore part of the sail is still flapping. You can also see this in the photo and figure *b* of the drawing. If you now pull in the sheet still further, the part of the sail that catches the wind and billows out becomes steadily larger, till the sail billows out completely. The boat is now going at its fastest (figure *c* of the drawing).

Backwinding the luff of the sail.

If you steer the bow of the boat into the wind (this is called luffing up) you see that the fore part of the sail backwinds. In order to go as fast as possible again, you must pull in the sheet to the point where the luff of the sail does not backwind (figures *d* and *e* of the drawing). If you luff up further you must continue to pull in the sheet of the sail so that it stays full at the luff.

If you now steer the boat away from the wind (this is called bearing away), you will see that the sail continues to billow out. Even when you ease the sail more, it still keeps billowing out. Now it becomes clear that the boat is fastest when you

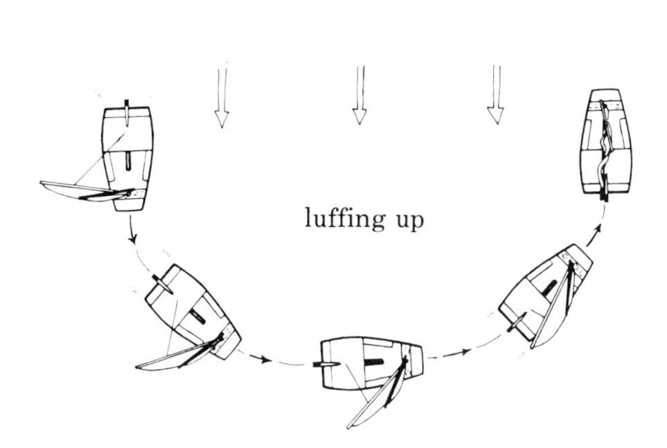

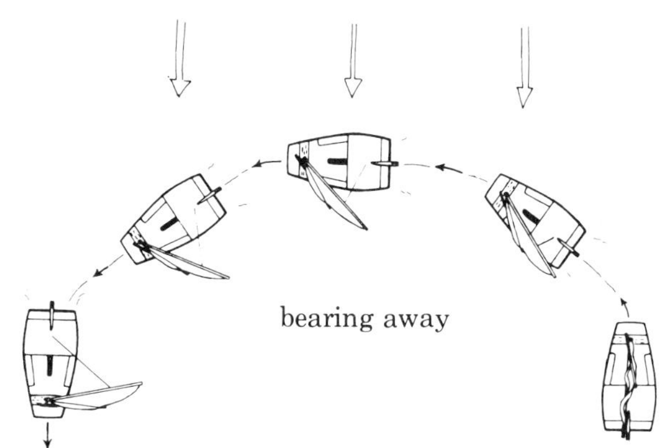

ease the sail as much as possible, but always check that the sail is not backwinding at the luff.
A sailing boat cannot sail directly into the wind. If you want to try you will find that the sail flaps around over your head, and the boat comes to a standstill and may even be blown backwards. You can sail in any direction except into the wind.

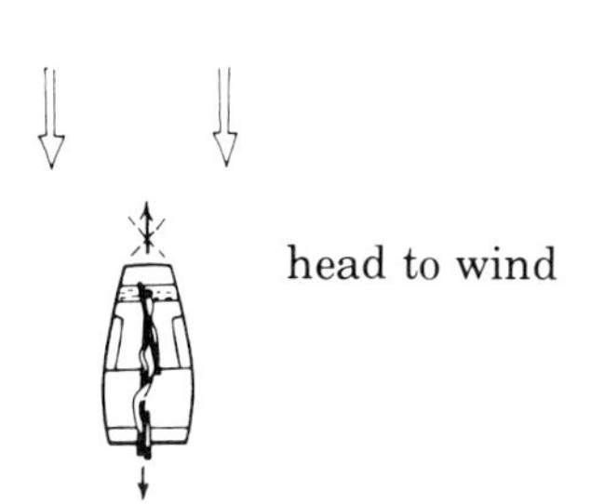

It is not easy to luff up and then bear away quickly, whilst tightening the sheet and watching the sail.

There is, however, an easy way of adjusting the sheet quickly. Tighten the sheet with one hand and then pass it to the other hand which is also holding the tiller. Tighten the sheet again with the free hand. Repeat this till the sail is fully tightened, all the time looking at the sail, and at where you are going. In this way you never have to put the sheet in your mouth to hold it, something that you see a lot of sailors doing.

Bearing away is easier. You must not let go of the sheet when easing the sail, but pay it out gently through your hand. In the following drawing you can see an overall picture of some courses that a boat can sail and the names that are used for them by all sailors.

Remember

- **The sail must always be as eased as much as possible, but the luff should not backwind.**

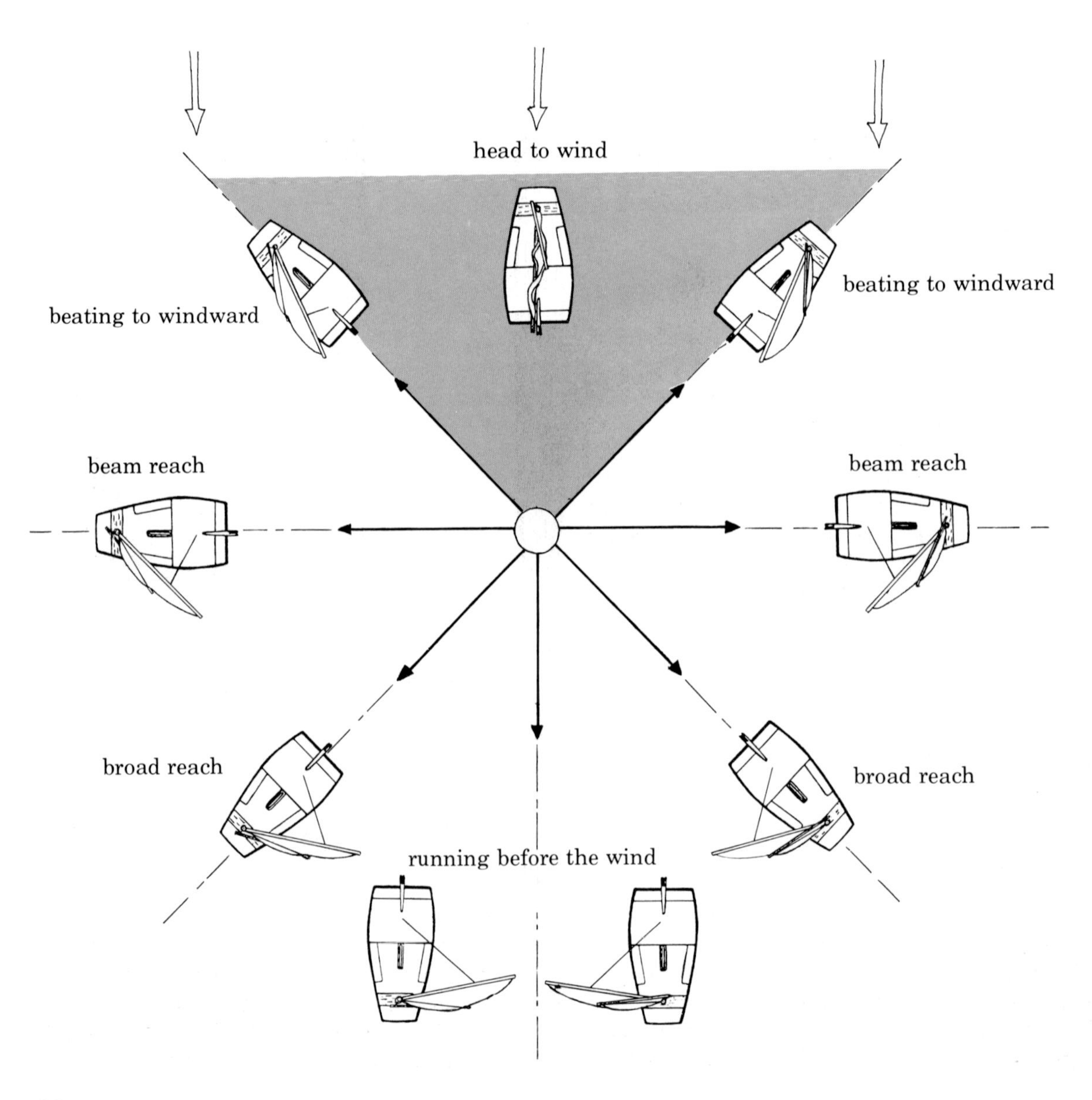
head to wind
beating to windward
beating to windward
beam reach
beam reach
broad reach
broad reach
running before the wind

7 Sitting out and the use of the tiller extension

It is important on a boat with a dagger board such as the Optimist, to sail it as upright as possible. (You can read about two exceptions to this in Chapter 21.1 *Weight trimming*.) This means that the boat should not heel to windward or to leeward. You may therefore have to move out more and more to windward as the boat heels to the wind. The first stage is to sit on the gunwale. If you still cannot keep the boat upright, you have to sit out. The technique of sitting out is described below. It places as much weight as possible to windward and you should be able to continue sitting out for some time.

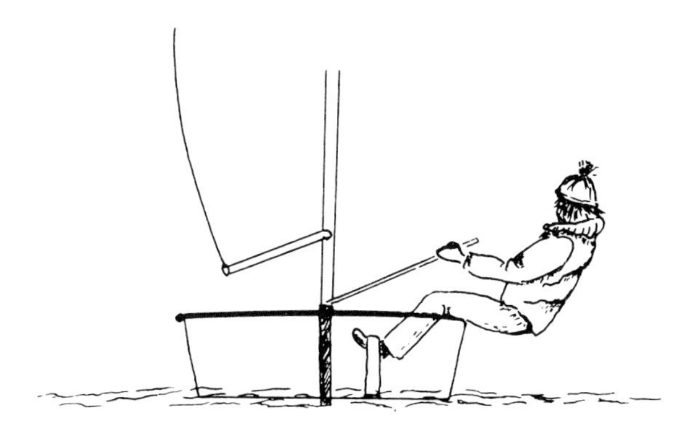

Sitting out using the toe straps and tiller extension

Sitting out

Place your feet under the toe straps which should be adjusted so that your thighs are on the gunwale when sitting out. In order to avoid any strain on your back, you should keep it rounded. You do this by putting your chin on your chest and pressing your buttocks tightly together.

If you are sitting on the gunwale or sitting out you can no longer reach the tiller, so you must use the tiller extension. There are many ways of gripping this, but the correct technique is shown in Photo 1. Your wrist and lower arm will be completely relaxed as you steer.

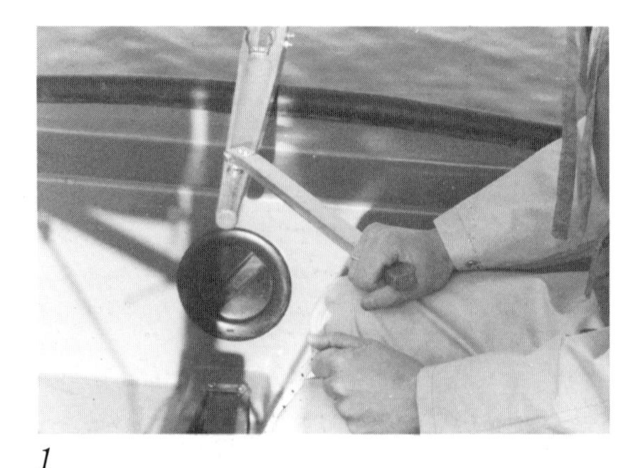

1

2

When it blows harder and the rudder pulls more it is better to position the tiller extension at your side but still keep the same hand grip as you can see in Photo 2.

The method of holding the tiller extension shown in Photo 3 is incorrect and will cause cramp. You will not be able to hold it like this for long, especially when it is blowing hard.

3

A
The sheet and tiller extension are held in the same hand.

The correct procedure for tightening the sheet when you are using the tiller extension is shown in Photos A, B and C.

B
Your spare hand grabs the sheet near the block and at the same time your steering hand is raised. (This can be done without changing course.)

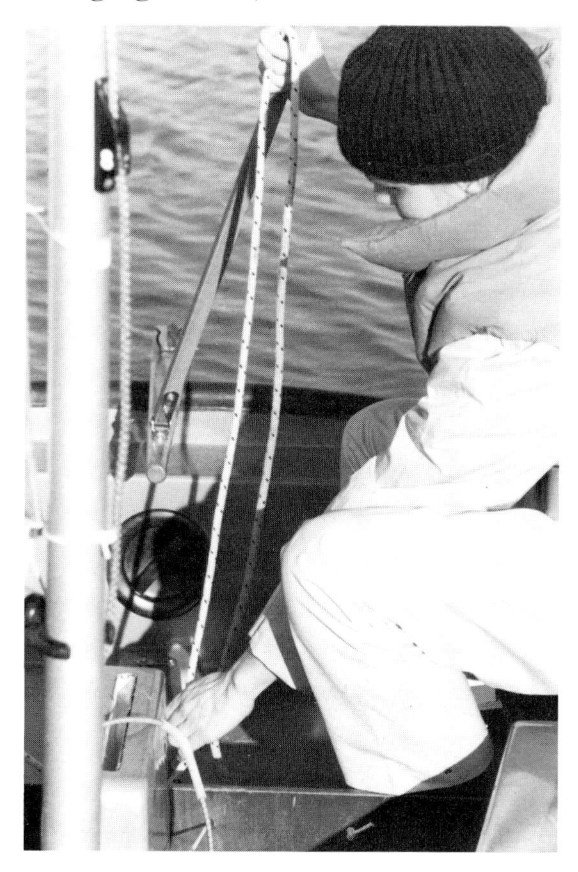

C
Release the sheet from your steering hand as you lower the tiller extension. You now have the sheet in one hand and the extension in the other. It may be necessary to repeat the sequence if the sail needs to be sheeted in further.

8
Turning through the wind

Imagine that you are sitting in boat A and you want to sail after boat B. Boat A is sailing with wind abeam, with the sail over to port. You can turn in two directions to come behind B.

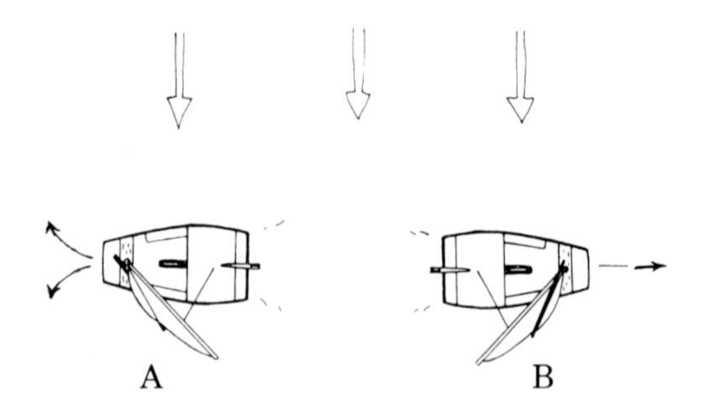

8.1 Going about

If you turn to starboard the boat will luff up. If you keep on turning, the sail starts flapping and the boat comes head to wind. If you now turn still further, the sail is blown to the other side. In this way the sail changes sides slowly, flapping as it does so. This turn is called 'going about'. As the sail changes side, you must also move to the other side.

You must always sit opposite the sail. Before you turn, first go down on your knees. Then push the tiller smartly away from you. You now move to the other side, as the boom swings across, but in the opposite direction. Thus: when the boom comes inwards, you also move in. You should be in the centre of the boat at the same time as the boom and when it has crossed to the new side you should be opposite. Throughout the turn you should continue to look ahead and keep hold of both the sheet and tiller. When you move across you should swap tiller and sheet hands.

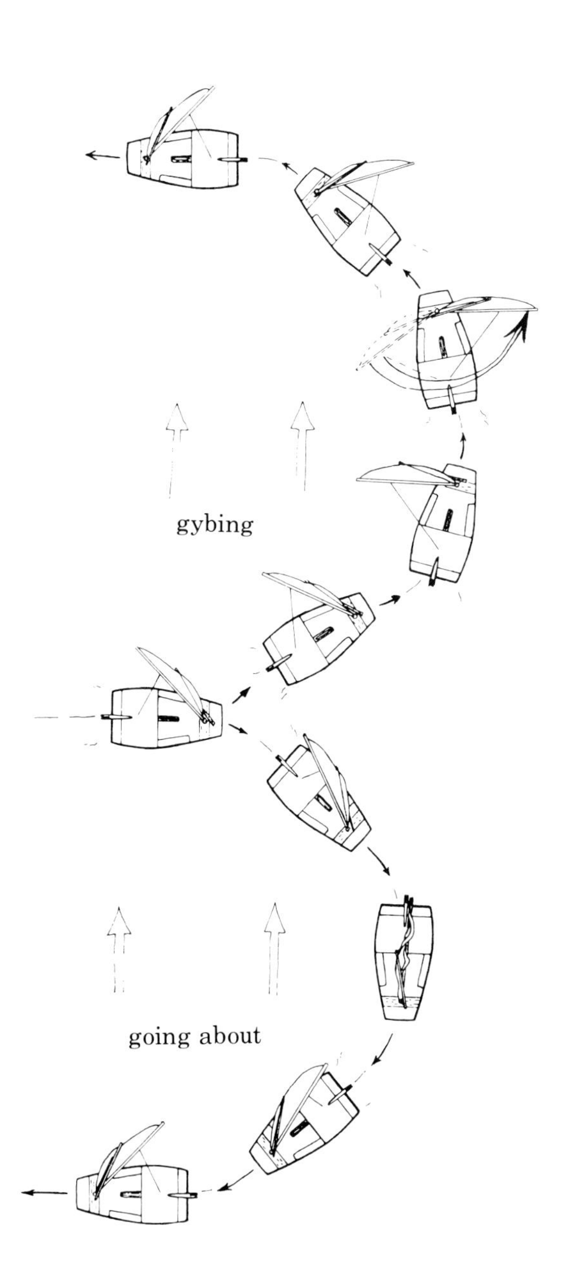

8.2 Gybing

If you turn A to port she will bear away. If you ease the sail the boat will eventually be running before the wind and if you turn still further the sail will suddenly change sides very quickly. This is called gybing. When gybing you must change sides as you would when 'going about'.

When transferring your weight, you must be more careful than when 'going about'. The sail comes over very quickly and immediately catches the wind. You can forestall a sudden gybe by pulling the sail in a little with the sheet, grasping it between the boom and bottom blocks. You must do this just before the sail gybes itself. Again, during a turn you must keep looking ahead and also keep hold of the sheet and tiller.

8.3 Going about and gybing with the tiller extension

When going about or gybing, turn the tiller extension backwards in an arc passing over the tiller itself. If you turn it forwards it often becomes tangled with the sheet. You will need to practise before you can gybe or go about quickly using the tiller extension.

9
The dagger board

Sticks, folded paper boats and so forth always blow along on the water with the wind. With a sailing boat that is luckily not so. The fact that you can sail with the wind abeam or even to windward is due to the dagger board. In order to try understanding this try pulling a plank through the water. With the narrow edge forward it goes well but with the wide side forward it takes great effort.

Now look at the drawings. The boat with the dagger board down is not pushed too far to the side or making leeway as it is called. This is because the dagger board resists sideways movement underwater. In a forward direction it goes well since only the edge needs to cut through the water.

When the boat runs before the wind the dagger board is not necessary. You can take it up and the boat will then sail faster.

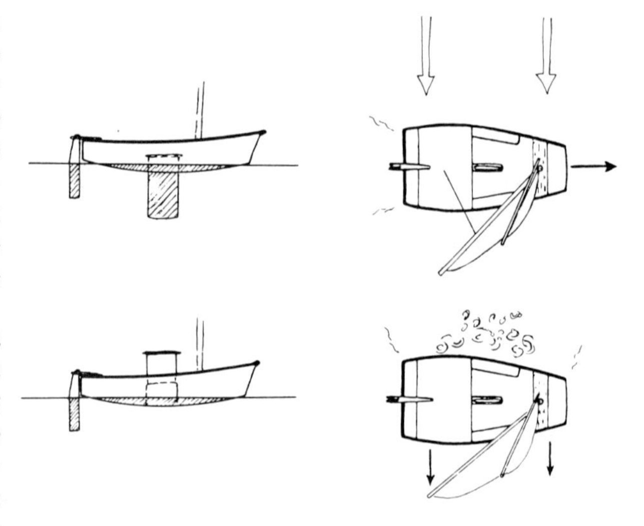

An elastic shockcord prevents the dagger board slipping down when running before the wind.

You have now seen that with wind abeam the dagger board must be almost completely down and when running before the wind, almost completely up. There are naturally many more points of sailing, each with its own dagger board position. So when you alter course you not only trim the sail, but must also adjust the dagger board. Sail to windward with the dagger board completely down and before the wind with it almost completely up. The other points of sailing have intermediate positions.

Remember

- **The dagger board prevents drifting to leeward.**
- **When altering course, remember to adjust the dagger board.**

10
Tacking and beating to windward

10.1 How to sail to a windward point

You have already read that you cannot sail directly into the wind, but it is possible to reach a windward point by sailing alternately to port and starboard. You do not sail with the wind on the beam as described in chapter 7, but sail as close into the wind as possible. This zig-zagging is called tacking or beating.

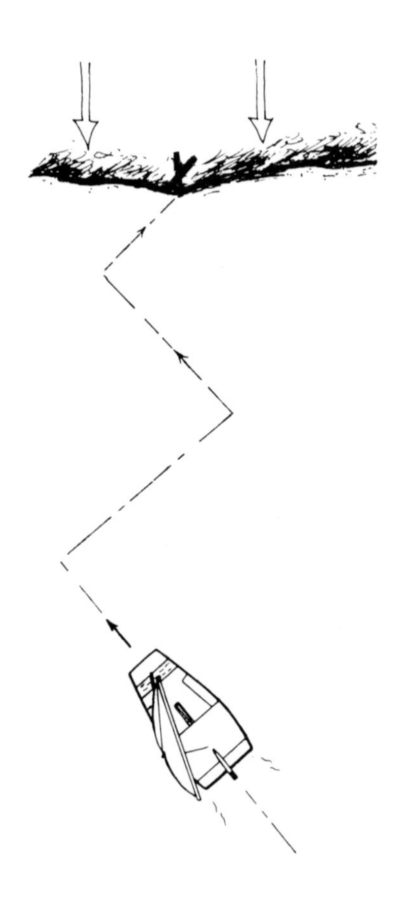

10.2 Bearing away from being head to wind

Sometimes if you do not steer very carefully or go about too slowly, the boat may come head to wind and stop. The sail will flap in the centre of the boat and you may start going backwards (or making sternway).

What can you do to stop this and continue on your way? Let the boat calmly make a little sternway. Leave the sheet completely slack and push the tiller in the direction that you wish to point. Sit on the opposite side to which you are pushing the tiller. When you are lying with the wind abeam pull the tiller towards you and then tighten the sheet slowly. Find the right course and the correct sail trim and continue on your way again.

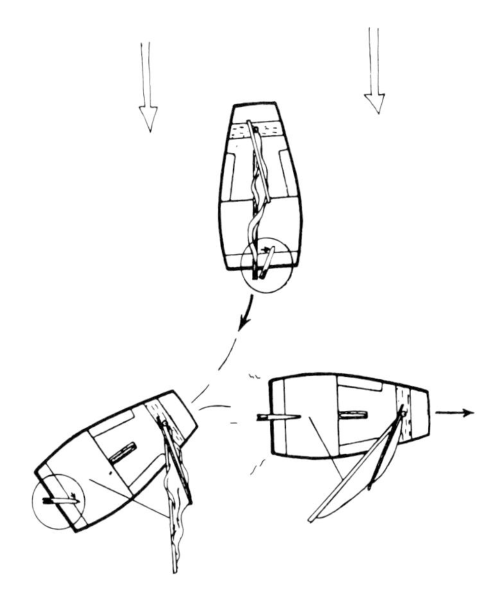

10.3 Sailing close-hauled

To reach a windward destination as quickly as possible you must sail as close to the wind as you can without losing the wind from your sail.

In the illustration, boat 1 is sailing to windward but wants to sail closer to the wind. The sheet is tightened to a point where the boom is over the aft corner on the leeward side of the boat (See boat 2). After that the helmsman luffs a little and looks at the luff of the sail. At a given point the sail will backwind (see boat 3) and the boat will slow down since there is now less wind driving the sail. The boat will heel slightly to windward; the helmsman has sailed too close into the wind. He will now have to bear away so that the whole sail fills again (See boat 4). He will feel the boat come upright again and gather speed. Boat 4 is now sailing close-hauled to windward.

Sailing close-hauled seldom means sailing a straight course because every time the wind changes strength or direction you must alter course to keep close to the wind.

Altering course is achieved through small movements of the tiller. The sail stays constantly in the same positon.

Sailing to windward is almost a sport in itself. You are always concentrating hard to sail fast and make progress. You need all your wits about you. You look across the water to see the gusts and then to the luff of the sail. You hear that the boat is going faster or more slowly. If you sail too close then you feel that the sheet and the rudder pull less. The boat then heels over to windward. If you sail too far off the wind then the boat will just heel over more to leeward and the sheet and the rudder pull more. In short, it is not a time to fall asleep.

The best way to 'go about' when beating is to sail from close-hauled on one bow to close-hauled on the other. This is called tacking.

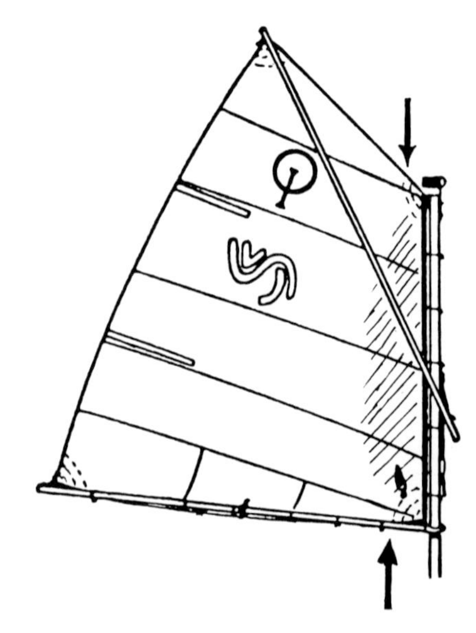

The part of the sail that you must watch when sailing close-hauled.

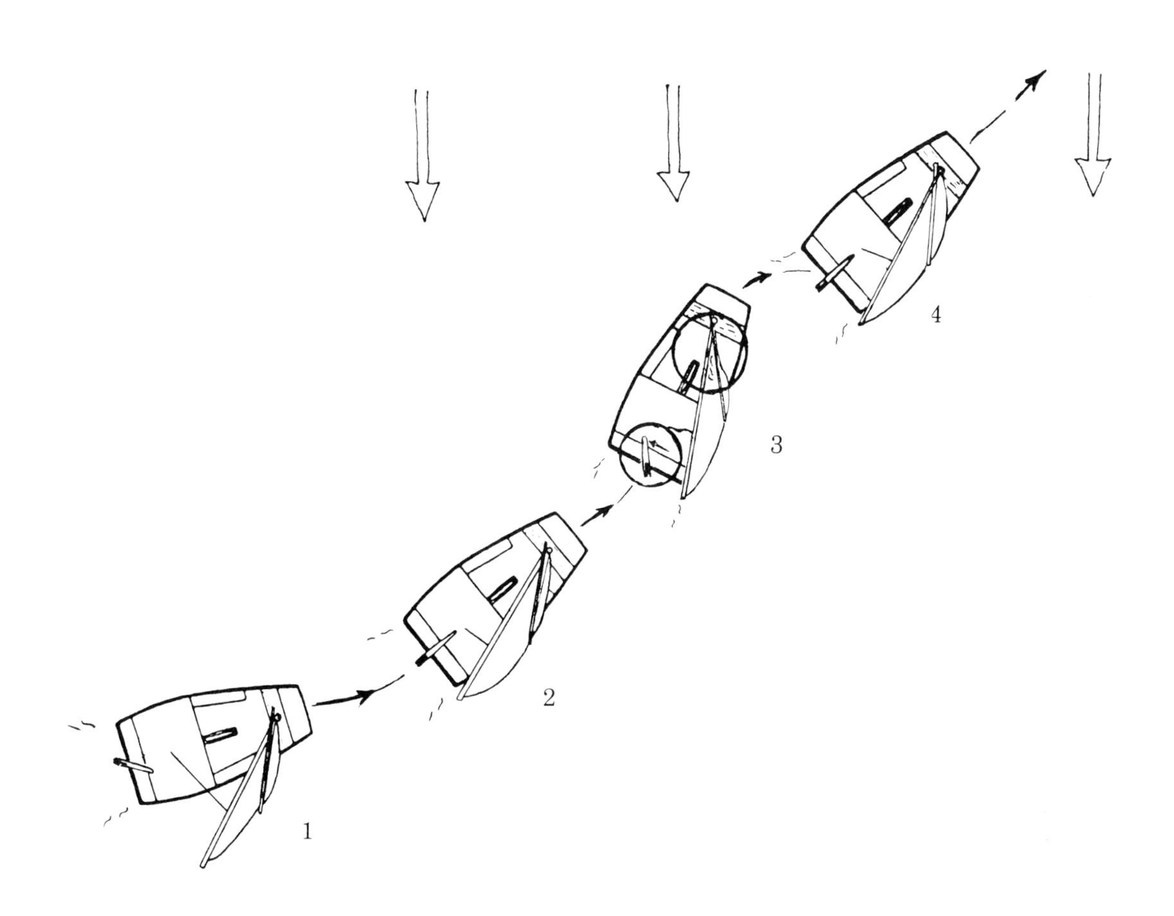

Remember

- **A close-hauled course is seldom sailed in a straight line.**
- **The sail is always sheeted in.**

Boat 3 is sailing too close to the wind.

Boat 4 is sailing close-hauled correctly.

10.4 Lay lines

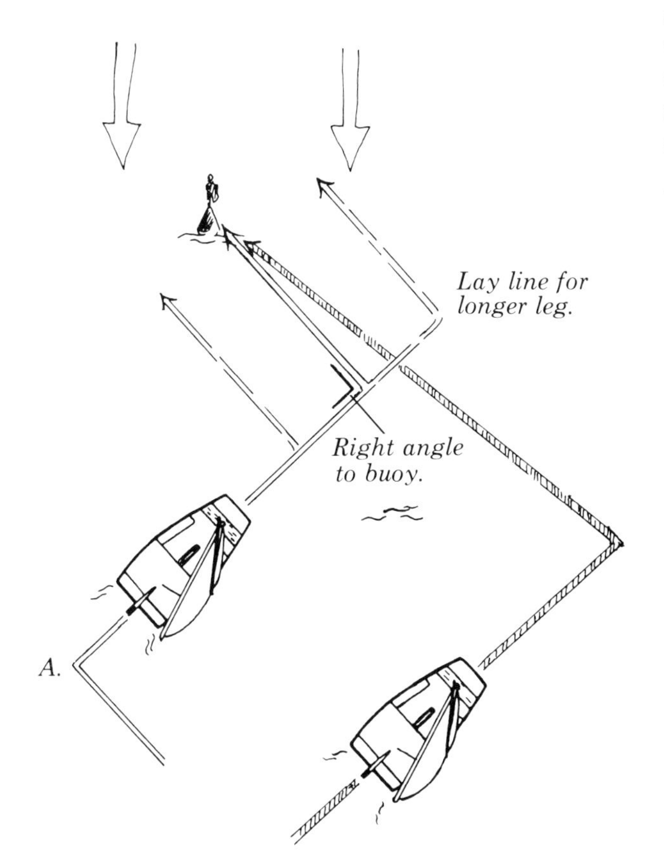

Imagine you are tacking as in the illustration and you have just gone about at A. You must tack once more in order to round the buoy. How do you decide when? In the illustration you can see that the windward course to starboard is approximately at right angles to the windward course to port. You can use that to determine when to go about.

When you see that the buoy is at right angles to you, sail on for a little and then go about. You will then see that you are on the correct course for the buoy, or the lay line. It is important to sail on a little after the buoy is at right angles, to counteract the leeway that is made when sailing close-hauled. You would probably not reach the mark if you tacked at right angles to it. So you should also remember that the further away from the buoy you are when you tack, the greater the margin that you should allow for leeway. Also, if the wind is strong and there are many waves, allow for even more leeway.

Remember

- **To tack on the lay line sail a little further than a right angle to the buoy.**
- **Compensate for leeway.**

11
Heaving to, lying hove to and controlling speed

In order to stop with wind abeam or closer to the wind you must be able to heave a boat to. This is possible by seeing to it that the sail catches no wind and flaps. When sailing with the wind abeam or closer to the wind, you only need to let out the sail, but on a downwind course this is impossible.

There is always a knot in the end of the sheet. Therefore it is not possible to let the sail swing ahead of the mast, so in order to heave to on downwind courses you will have to alter course into the wind. The sail can then flap free and you will stop. The boat can also be steered right into the wind. In this way the boat heaves to fastest, but once you are motionless it is difficult to make forward speed again. How you do that is stated in chapter 10.2 *Bearing away from being head to wind.* In order to remain motionless as easily as possible, you should adopt the following technique. Put the boat close to the wind with the sheet eased off so that only the leech catches the wind. Let the

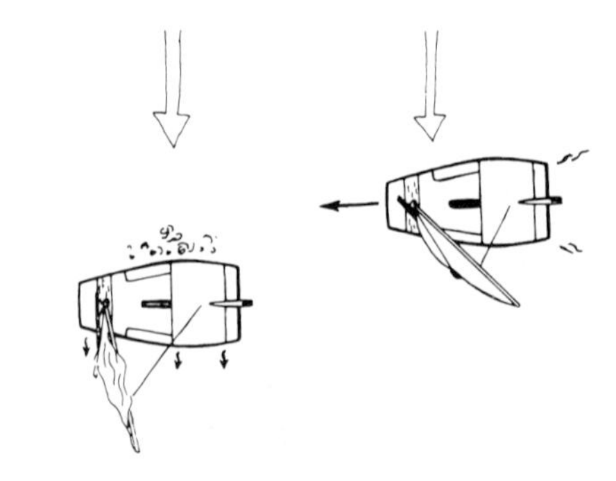

Heaving to when wind abeam

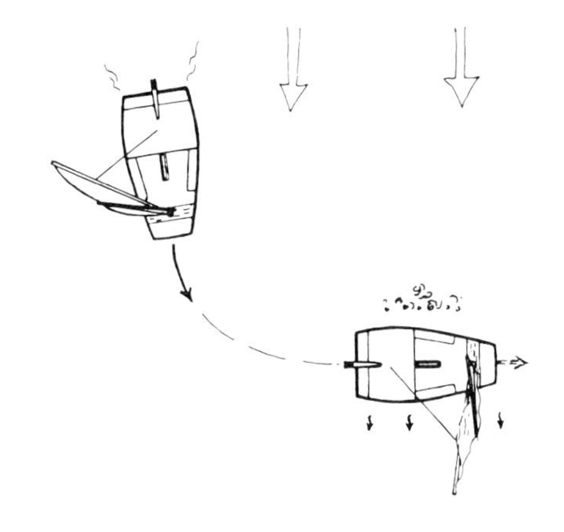

Heaving to when wind behind

boat heel over a little to windward. In order to make sure that the boat remains motionless, carefully waggle the rudder (move the tiller to and fro).

In order to gain speed again, you only have to sheet in. You can also partially sheet in if you wish to go more slowly. If you let the whole sail catch wind at once, then the boat will drift quickly to leeward.

Remember

- **Keep wind in the leech only.**
- **Stay on a course close to the wind.**
- **Heel to windward.**
- **Waggle the rudder to and fro.**

12
Coming alongside and casting off from a windward shore

The wind blows from a windward shore and towards a lee shore. When coming alongside it is important to remember the difference. In this chapter we shall talk about the windward shore.

There are two ways to approach a windward shore properly, safely and quickly.

One is the bows-on approach. Sail along beside the shore or jetty that you are aiming to come alongside and then heave to quickly by steering into the wind and letting off the sheet. Through the quick turn the boat loses way immediately. If you are still going too fast push the sail against the wind (set the sail aback). You can make a bows-on approach from any direction.

Another method that takes a little longer is called the windward approach. The speed of the boat can be carefully controlled during this manoeuvre. The windward approach means that you sail on a windward course to your landing point. Before you get too close to the shore ease the sail so that only the leech

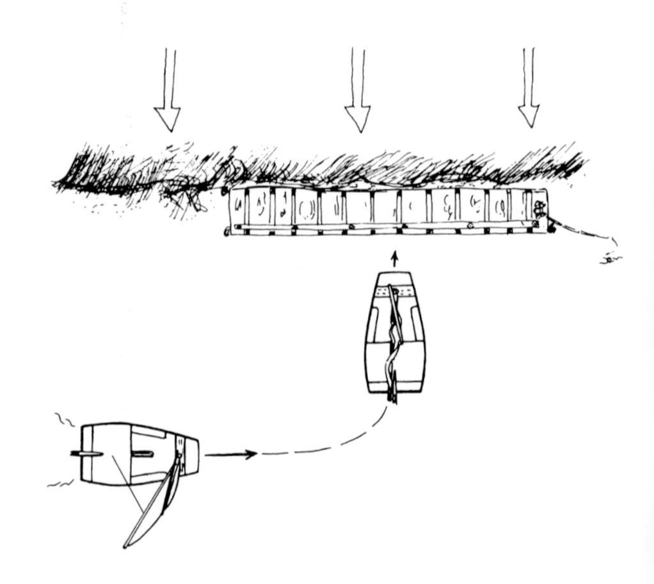

Bows-on approach.

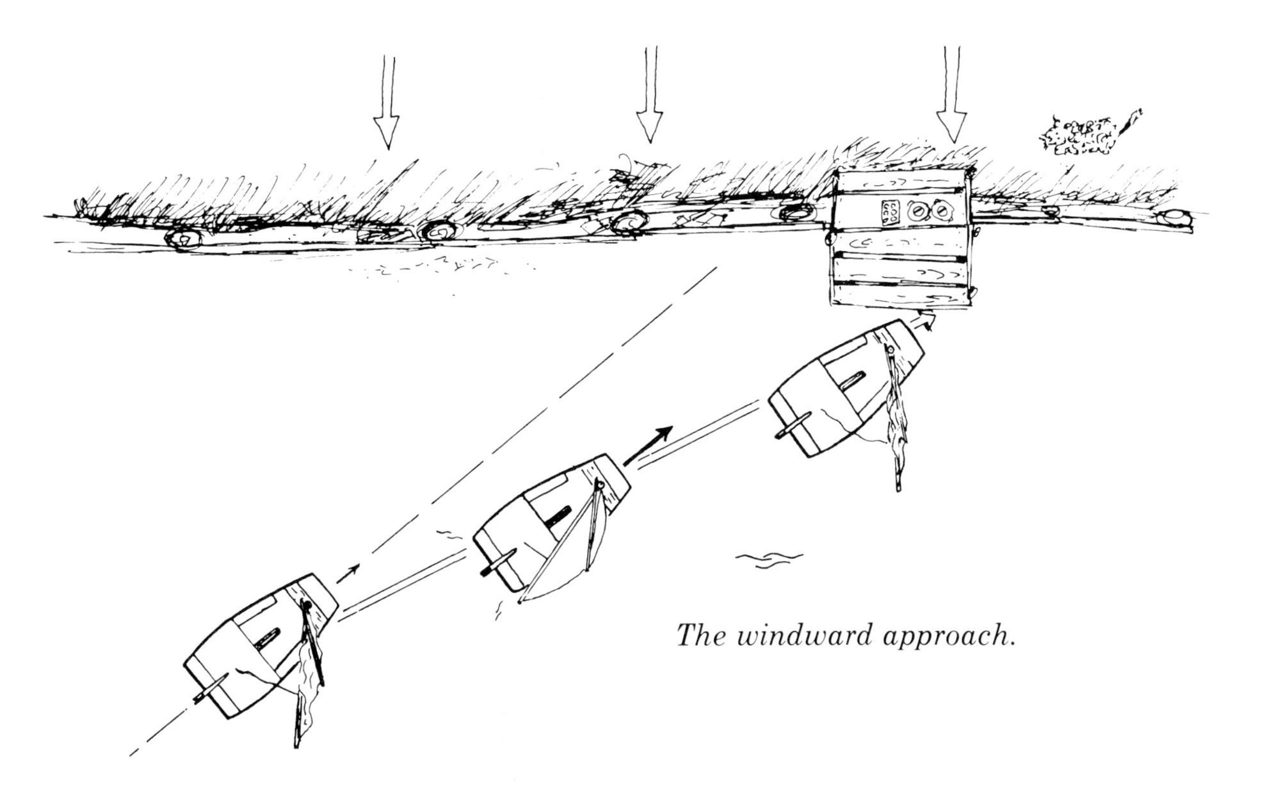

The windward approach.

catches wind. By doing this you keep a little speed on and will drift less to leeward. Keep the boat upright as this also lessens drift to leeward. Just before the shore, ease out the sail completely.

If the boat stops short of the shore, sheet in the sail a little. To prevent making leeway you should do this slowly.

When approaching a windward berth or landing point it is a good idea to steer slightly to windward. You can then allow the boat to drift to leeward and arrive at your intended landing.

The advantage of the approaching to windward technique is that the manoeuvre is controllable and you also have plenty of time to correct any misjudgements. It is also, therefore, safer. The disadvantage is that it takes longer than the bows-on approach. You should practise the bows-on approach a few times away from the shore to prevent denting your boat.

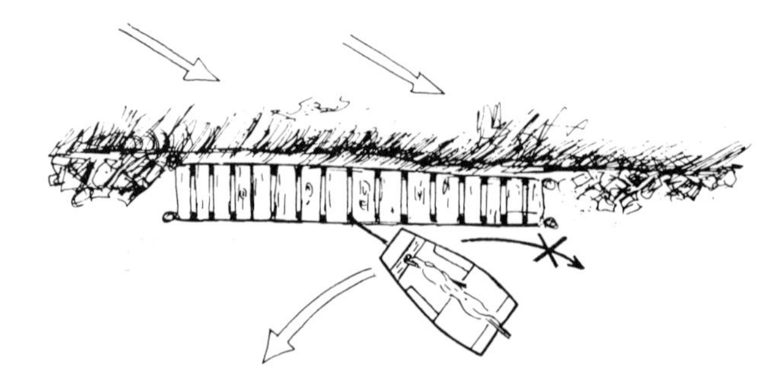

Casting off on the correct bow.

To cast off, give the boat a push in the right direction so that it turns the right way. Before casting off, take a good look around to see that no other boats are approaching. It is best to sit on the side of the boat that will be to windward once you assume your desired course. As soon as you have cast off push the tiller in the direction of your intended course and when the boat gains speed pull the tiller towards you and sheet the sail in.

It is important to cast off on the correct bow when the wind is at an angle to the shore. The boat in the illustration is casting off on the correct bow. If you were to cast off on the other bow you will sail in a long curve and may scratch the boat.

Remember

When approaching a windward shore

- **Keep the boat upright.**
- **Slow down by only using the leech of the sail.**
- **Ease the sheet to adjust your approach speed.**

When casting off from a windward shore

- **All clear?**
- **Sit on the eventual windward side.**
- **Work out the angle of the wind from the shore.**

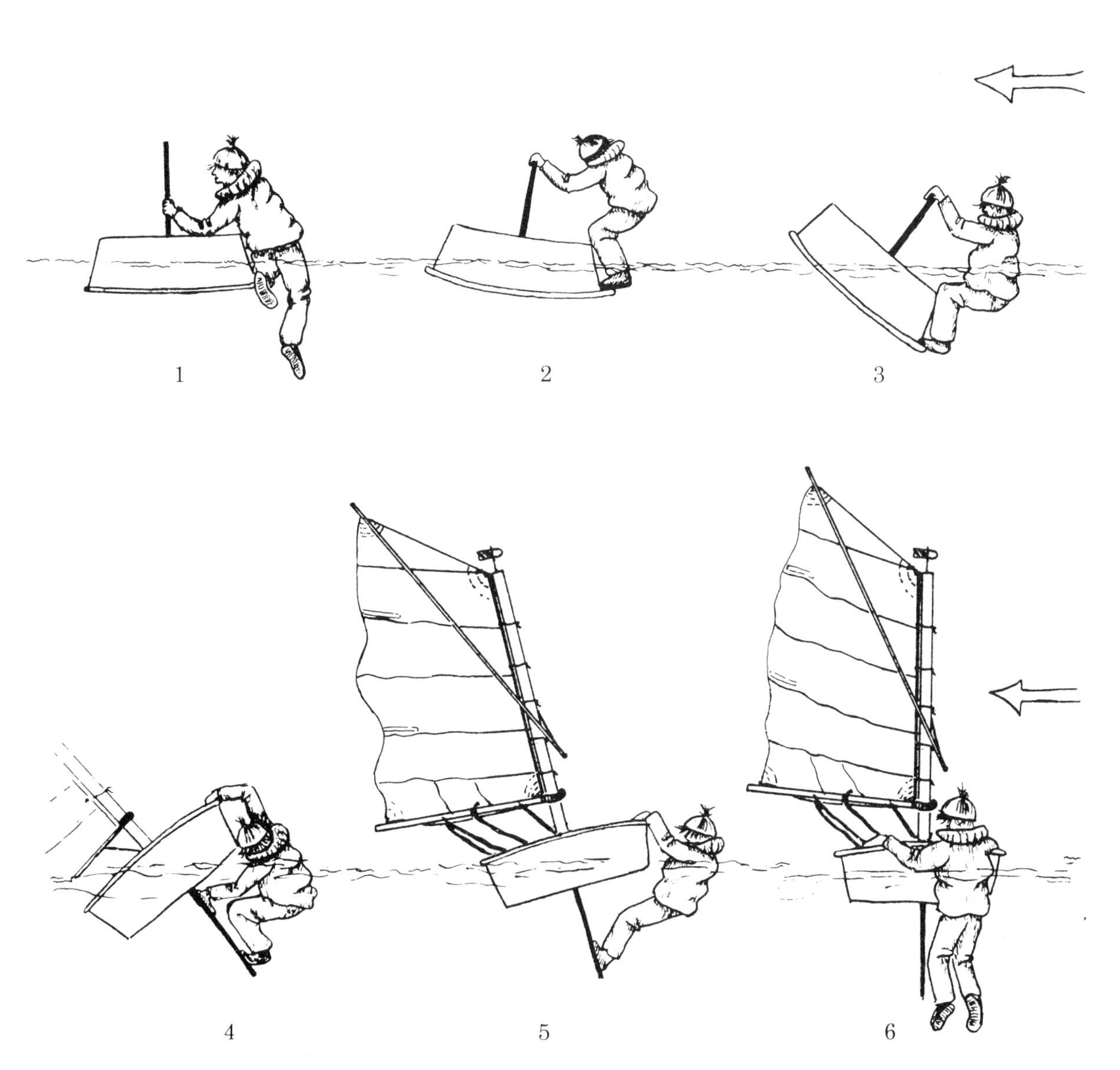

Righting the boat (See following chapter)

13
Righting the boat

Splash ... and there is the boat capsized and you are swimming in cold water. What now? Right the boat but never swim away from it. In deep water or at sea, the boat may drift quite fast, so get a grip on it as soon as you can so as to prevent it drifting away from you.

Now climb on to the hull on the windward side; hang on to the dagger board with your hands and push on the boat with your legs (See Illustrations 2 and 3 on previous page). The boat will then right itself slowly and as soon as the dagger board is almost in the water, climb on to it (See Illustration 4). Hang on to the gunwale and push against the dagger board with your feet (See Illustration 5), the boat will right itself completely.

Pull yourself round to the stern and climb aboard (See Illustration 6). Look to see if everything is still in position and bale the boat dry.

It can sometimes be difficult to right the boat; if so, here are some useful hints: If you climb on to the hull from the leeward side the boat will often capsize again. Look at the illustration and notice how the wind will first blow against the hull and then the sail. The boat will blow straight over again as a result.

The sheet can present another problem if it is still cleated. As soon as the sail comes out of the water it will fill with wind and the boat will either sail away or capsize again.

If you right the boat too quickly it will ship a great deal of water, but if you right it more slowly you will find less water to bale out.

It is sensible to practice righting the boat in a light wind, so when it happens in a strong wind you will feel much more confident.

Finally: When you capsize you will appreciate the careful preparation that made sure your lifejacket was safe and secure and that the caps were secured to the boats buoyancy tanks.
Remember - Check everything before going afloat.

14 Paddling

It is not always possible to sail your boat. You may meet a bridge that is too low to sail underneath or the wind may die down when you are far from shore. The sail may tear or a sheet may break. In these sorts of situations you may have to paddle the boat to safety. Often when out sailing you do not have oars and a rowing seat with you.

To help you paddle efficiently you must take off the rudder, then move forward in the boat, with the idea of sitting on the mast thwart. Take out the mast and roll up the spars and sail. It is best to sit forward in the boat since you can then steer more easily. If you paddle on the right-hand side of the boat it will go to the left. Paddle on the left of the boat and it will go to the right.

You can apply more power if you hold your hands apart from each other on the paddle.

To paddle a straight line make a few strokes on the left and than a few strokes on the right.

You will always have to make more strokes to leeward than to windward. Just look at the drawing. The wind is blowing **Snoopy** to port. In order to sail a straight course the paddler of the boat **Snoopy** will have to paddle more strokes to port, or to leeward.

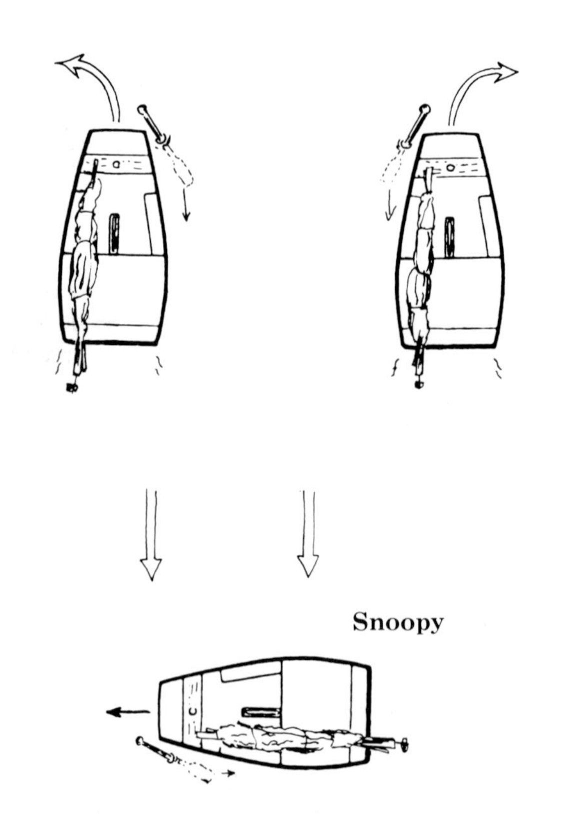

15
Rowing

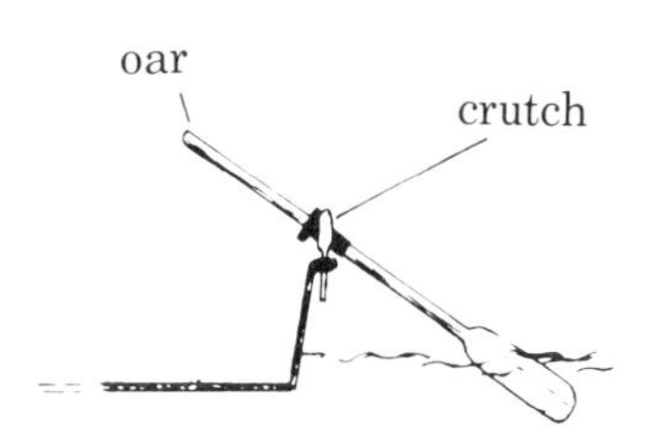

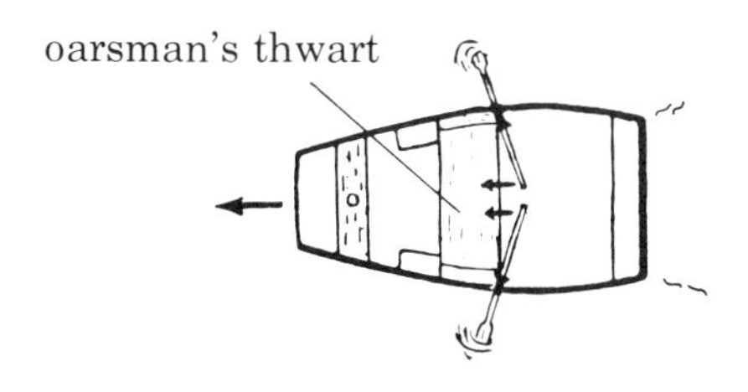

The Optimist is not only an excellent sailing boat, but you can also row it. To do this you need an oarsman's thwart, two crutches and two oars. In most Optimists everything is attached as in the drawing.

When rowing, you sit on the oarsman's thwart, with your face towards the stern. Grasp the oars at the inner ends. Never stick the blades completely in the water, since they may shoot out of the crutches. The technique of rowing is as follows:

- Pull the blades out of the water and push the oars as far as possible away from you.
- Stick the blades in the water and pull the oars with stretched out arms towards you. (In this way you apply the power with your whole body, instead of only with your arms).
- When the oars are by your body, pull them out of the water and repeat the sequence.

Steer by applying more power to one oar than the other. If you want to turn quickly pull one oar while pushing the other. By keeping both oars in the water and exerting a steady pressure against them you will be able to slow down the boat.

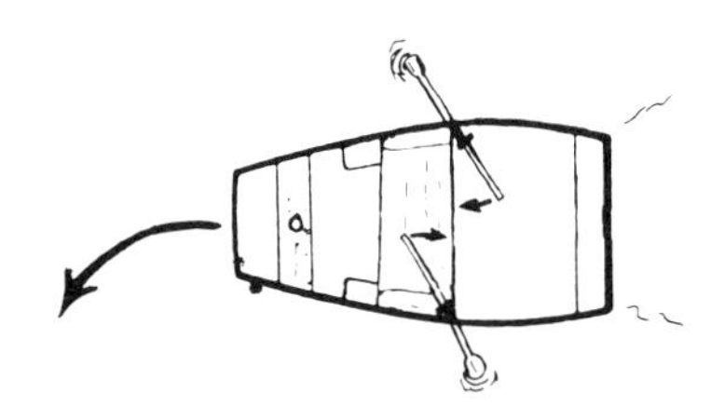

16
Taking a tow

Boats are generally towed when the weather does not permit sailing, sometimes because there is too much wind and sometimes because there is too little. The most important thing to remember when under tow is that the sail must always be taken down, otherwise the towing boat would only be able to steer into the wind. In addition, it will be safer as the boom won't hit your head. Also, there will be less wear and tear on the sail and you'll be able to see where you are going.

If you want a tow from a passing motor boat you should indicate this clearly. Do not shout, but wave in a friendly manner and hold your painter so that it can be seen clearly.

If there is no wind, take the sail down immediately you realise that the towing boat is approaching. If it is a windy day it is better to go alongside the towing boat (which should have stopped head to wind). It is easier to strike the sail pointing into the wind and also to pick up the tow. The towline should be about three to four metres long (12 feet). Pull the dagger board up to make steering easier and sit at the stern of the boat. Have a paddle at the ready.

When you want to drop the tow, you must tell the skipper of the towboat. He will then slow down. Thank the skipper and then let the towline be thrown into your boat. You could of course, stay alongside the towboat to insert the dagger board again and if need be hoist your sail.

Sometimes several boats can be towed by one motor boat. Good organisation on the part of the towboat skipper is very important in this, therefore always follow his instructions. Generally, all boats come alongside the towboat in turn. They strike sail and pull out the dagger board. The first boat goes and ties on to the end of the towline. Each time another boat is added, the towline is paid out a bit. You must keep the painter short so that the boat cannot swing out too far. Casting off of the tow is done from the rear. First the aft boat casts off,

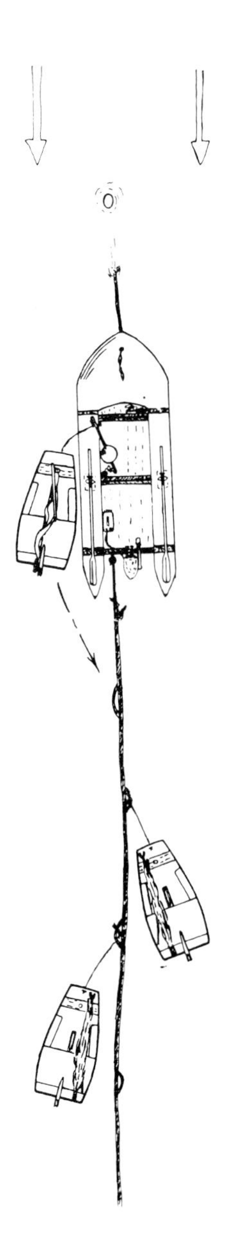

then the one forward of that and so forth. The towboat must be facing into the wind if you want to hoist sail whilst being towed.

Remember

- **Take the sail down.**
- **Take out the dagger board.**
- **Sit at the stern of the boat.**
- **Have the paddle ready.**
- **Steer to follow the towboat.**
- **After the tow, thank the skipper.**

17
Knots and hitches

Learning to tie knots is essential. A correct knot always holds fast and can be easily undone. Knots and hitches used in the Optimist are as follows:-

half hitch *two half hitches* *running half hitch* *two half hitches with the first one running*

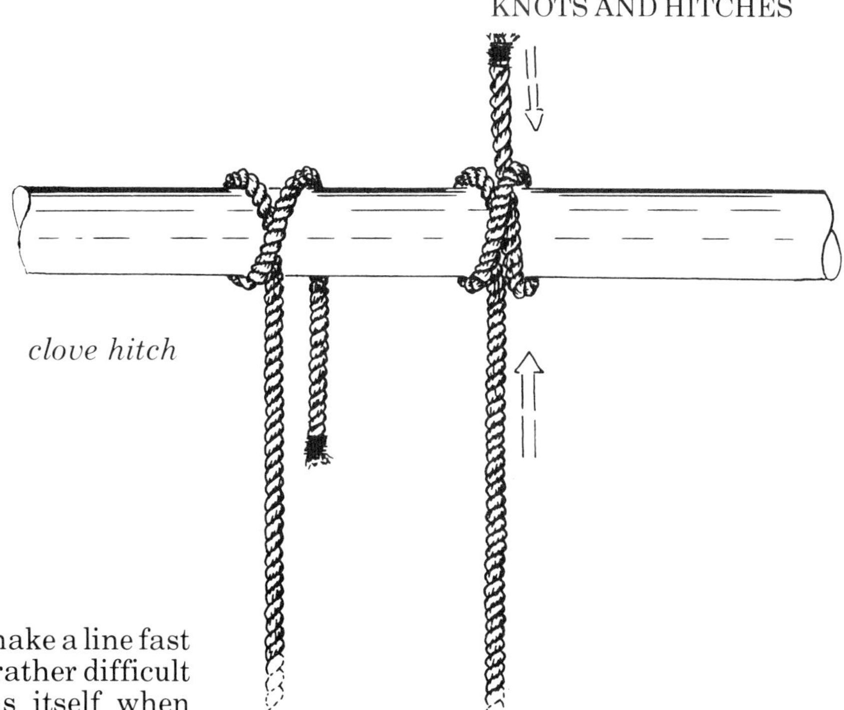

clove hitch

Two half hitches

This is the easiest way to make a line fast to an object. This hitch is rather difficult to undo, since it tightens itself when there is tension on it. It is easier to untie if you turn the end and at the same time push it back through.

Two half hitches, with the first one running

If you want to be able to undo two half hitches more easily you pull through the first hitch completely. You then make the second hitch with the loop. You can use this hitch to make a painter fast to a post.

Undoing it is quite simple. Press the second loop of the second half hitch up. Then pull the first half hitch out by the loose end.

Clove hitch

This is a rope crossed twice round a post. If you pull on both ends of the line, you only pull the hitch even tighter around the post. This hitch is used among other things to fasten the sheet to the boom; you should tie a figure of eight knot at the end securing the boom to prevent it slipping off. The hitch is also used to make the boat fast, but often with two half hitches added. Undoing a clove hitch is only possible if there is no tension on the hitch. You press the two ends towards each other with both hands. Then slip the two loops off the post.

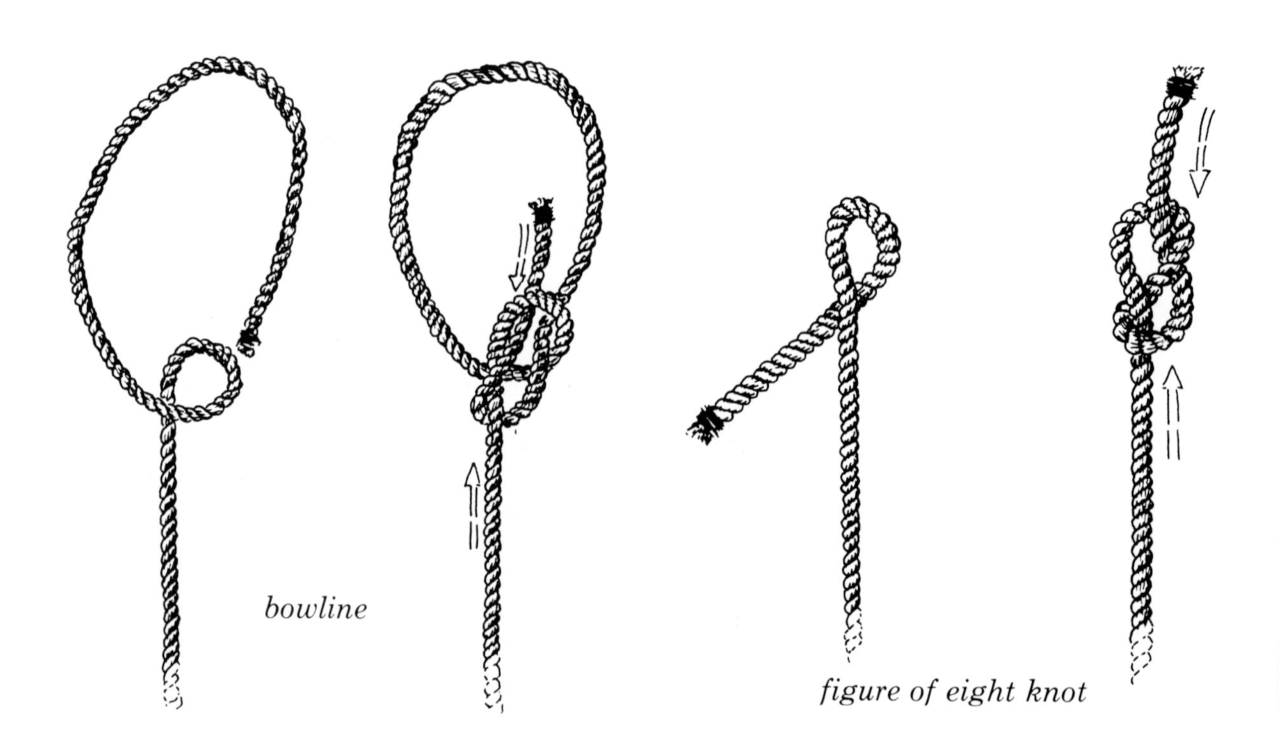

bowline

figure of eight knot

The bowline

The bowline creates a fixed loop at the end of a line. In the Optimist it is used to fasten the halyard to the sprit and the kicking strap to the boom. It can also be used to make the boat fast. Undoing this knot is only possible when there is no tension on the line. Then you can pull the fixed end back a little. The extremity of the line can then be easily pulled out of the loops.

The figure of eight knot

This knot has the shape of an 8. You always put the knot on the end of a sheet to make sure that it cannot fly out of the blocks.

The figure of eight knot cannot be pulled really tight because there are many bends in it and it is therefore easy to undo.

reef knot

Reef knot

This knot should be used to join two lines of about the same thickness to each other. It is easy to remember how you tie this knot with the following memory aid. You have a line in both hands. Put the end from your left hand over the end from your right hand. Pull through underneath, then right over left, pull through underneath and tighten. You use this knot to fasten the sail to the boom and the mast.

You undo it by holding one line in each hand and pushing them towards each other. It is easy to undo.

Frapping

When you stow the rigging you frap the sheet around it. That is done as follows. Put the sheet round the sail and through under itself. Fasten the end with two half hitches.

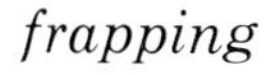

frapping

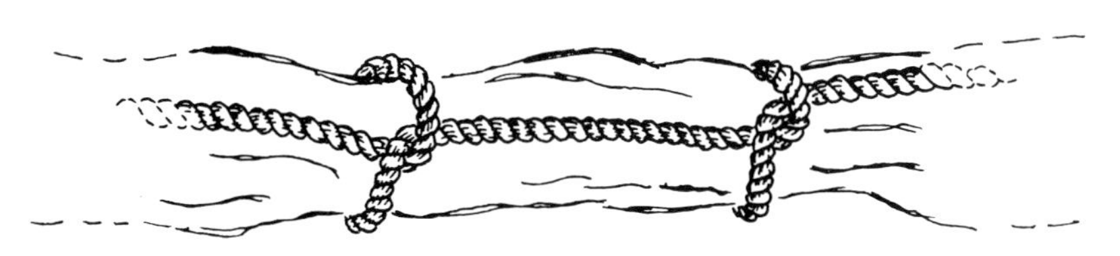

18
The rules of the road

Large cargo vessel

Small sailing dinghy

Wherever you are you must keep to certain rules; at school, at home or on the street. It is also like that on the water where rules help prevent collisions and accidents. All rules that apply on the water are known as 'The rule of the road'.

If you are not a very experienced sailor you will find it difficult to remember all the rules; it is hard enough to remember how to sail, but you should remember those rules which are likely to be needed as an Optimist sailor.

You should, from the outset be alert to danger, so keep a good look out. If you are heading straight for someone in your boat, give yourself plenty of time to take avoiding action.

The most important rules for the beginner to remember are:

1. You must always prevent a collision. If a boat should give way to you but does not, then you must alter course to avoid a collision.

2. Small boats must give way to larger vessels - because they cannot stop or turn quickly.

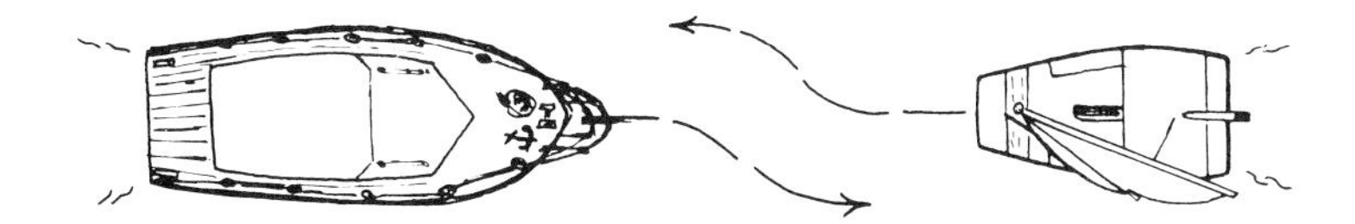

opposing courses
motor boat - sailing boat

3. If two boats (sailing boat, rowing boat, motor boat, canoe) **are sailing directly at each other,** they must **both take avoiding action** by turning **to starboard.**

Opposing courses
sailing boat - sailing boat

crossing courses *sailing boat - sailing boat*

4. For boats which **are sailing towards each other at an angle,** the following applies:
- the boat that has the wind to port must give way to the boat that has the wind to starboard.
- if both boats have the wind to port or wind to starboard then the following applies:

The boat which is to windward must give way. In short this rule says: Windward boat gives way.

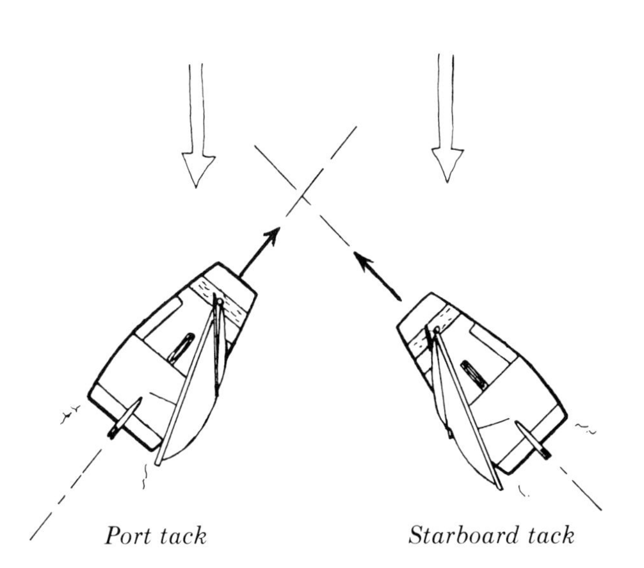

crossing courses
sailing boat - sailing boat

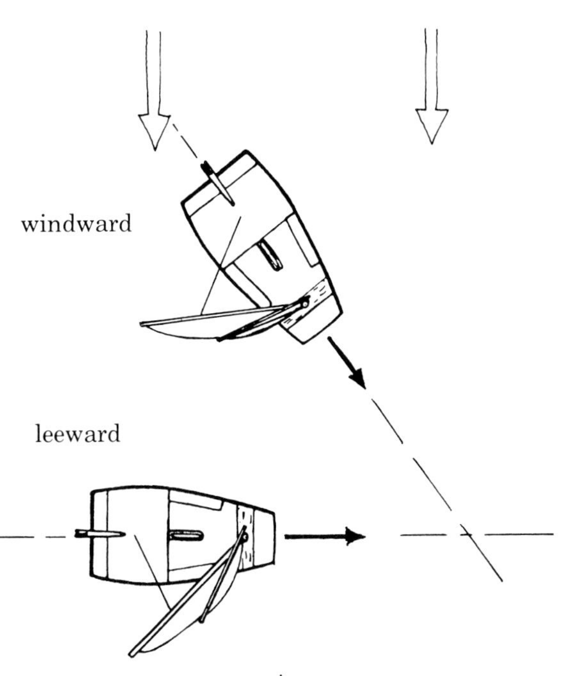

crossing courses

Crossing courses. Sailing boats and motor cruisers or rowing boats.
Rowing boats and motor cruisers should give way to sailing boats if they are approaching each other at an angle. (But in a confined narrow channel you should remember that boats should keep to the starboard side and you may have to give way on these occasions.)

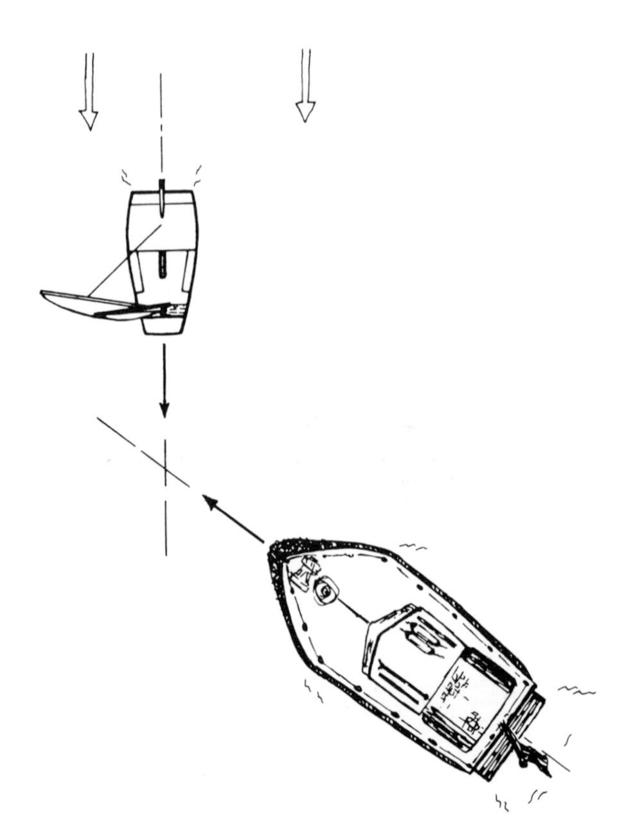

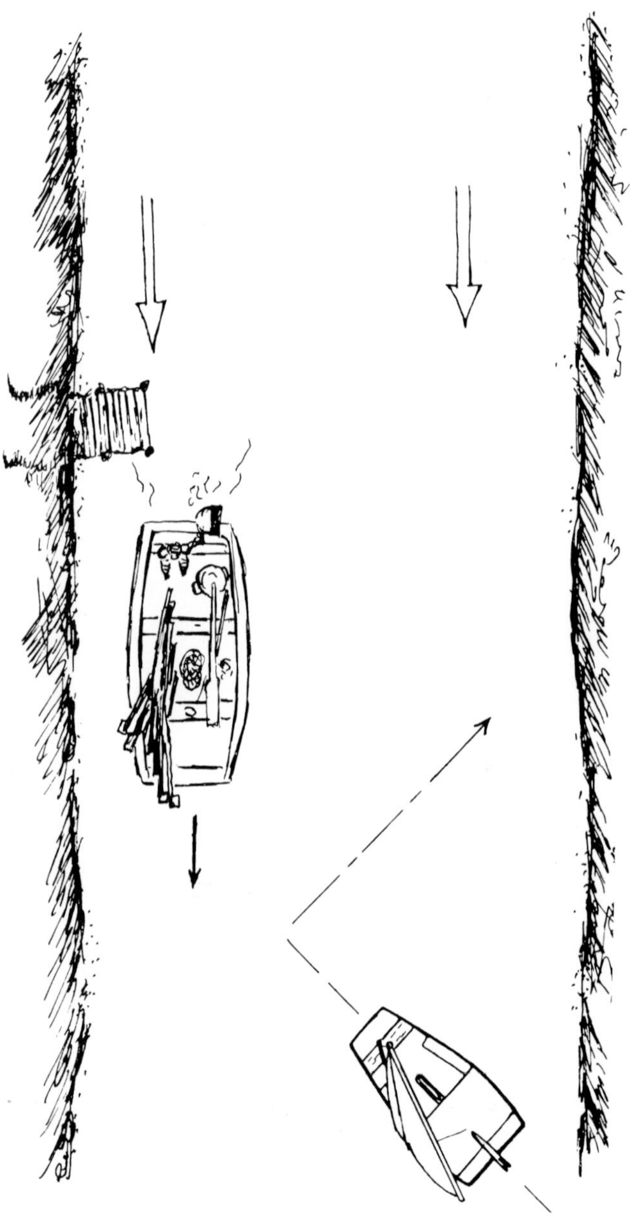

5. For boats which **are overtaking each other** the following applies:
- a boat that wants to overtake another boat must itself keep clear.

6. **Always wait before casting off until all vessels in your path are at a safe distance.**

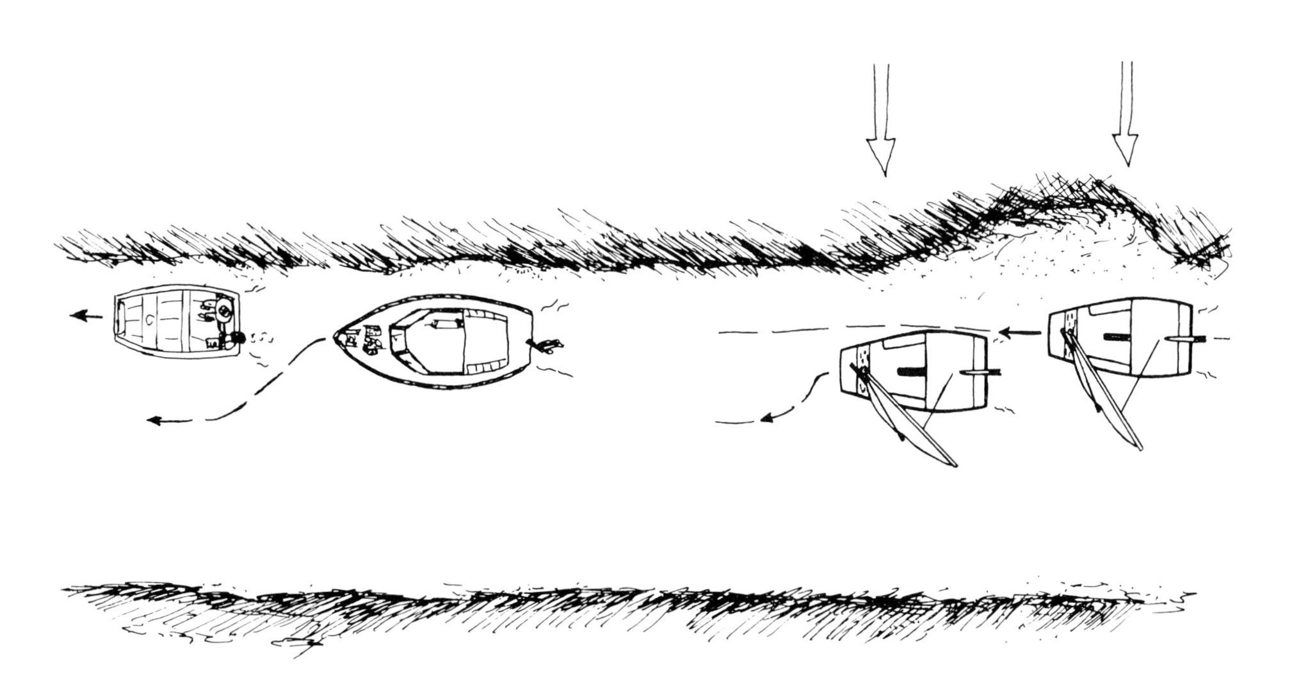

19
Gusts and changes of wind

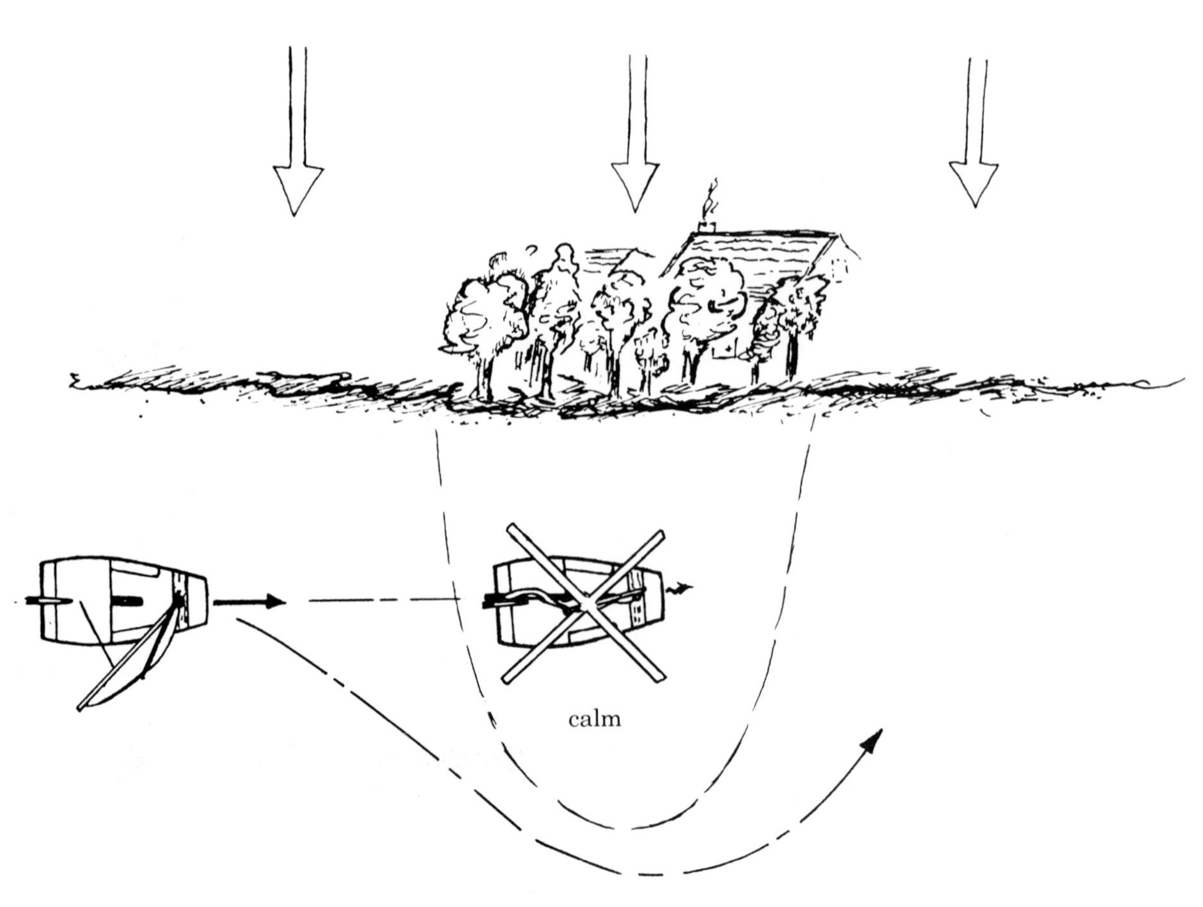

The wind is never constant in strength. In a gust there is more wind and in a calm less wind. When you look across water it is easy to see where gusts are and where there are calm patches. You recognise gusts by darker patches on the water, caused by many small ripples, and the calms by smooth water. When sailing keep a good lookout for them, because if you know in advance when to expect a gust or a calm, you can react in good time.

You often find calms along windward shores. Trees, houses and clumps of reeds shield you from the wind. Also when you pass a boat to leeward you sail through a calm patch. When this happens, the sail will start to flap and the boat will heel over to windward. If you are sailing close-hauled you should bear away. If you are sailing off the wind you should pull in the sheet. As you sail into more wind luff up again, or let the sheet out to continue on your correct course.

Sometimes you can bear away and sail around a calm patch. This keeps the boat sailing at a better speed.

In a gust the boat heels over more to leeward. You must prevent this by bringing your weight more to windward and luff up a little without losing speed. After the gust you bear away again. The advantage of this is that with each gust you gain a little distance to windward. If you don't want to gain distance to windward but speed, do not luff up, but just let out the sheet. After the gust tighten it again.

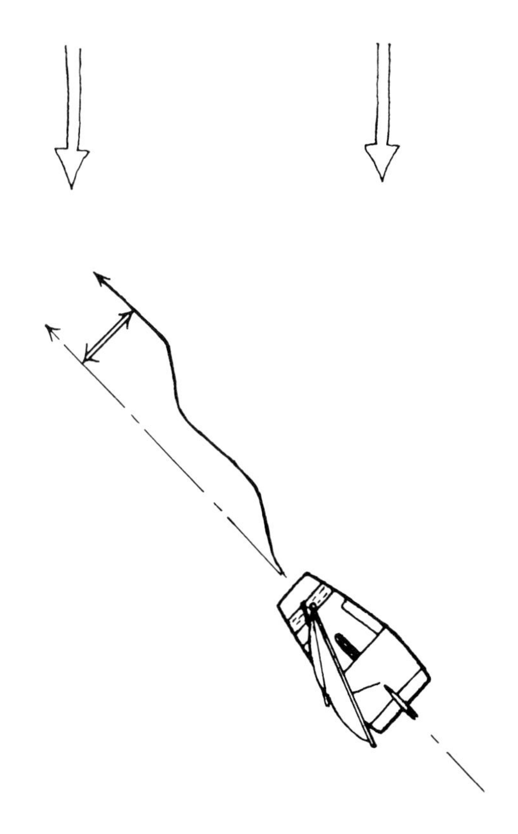

Gaining distance to windward in the gusts.

On more downwind courses you must also bear away a bit in a gust. This gains speed and you benefit from the gust longer. After the gust luff up a little again.

The wind is also not constant in direction, especially in light breezes when it changes direction quite frequently. If the wind direction moves aft, let out the sail. Conversely sheet in if the wind moves ahead.

Tacking on wind shifts

Sailing close-hauled is the art of sailing as close to the wind as possible without losing boat speed. If you are sailing to the limit and the wind changes you must take immediate action. You can see this in the illustration. Boat **Asterix** is headed too close to the wind and the sail is starting to backwind and flap which will cause the boat to lose speed.
The boat must bear away in order to stay close-hauled. Boat **Obelix** has the advantage of the wind change. The wind has moved a little aft or freed as it is called. Boat **Obelix** can therefore luff up in order to become close-hauled again and therefore sails a closer course to his windward objective than before.

For boat **Asterix** the wind change is disadvantageous. It must bear away and therefore sails further away from its objective. You can see this clearly in the illustration. Boat **Asterix** is sailing

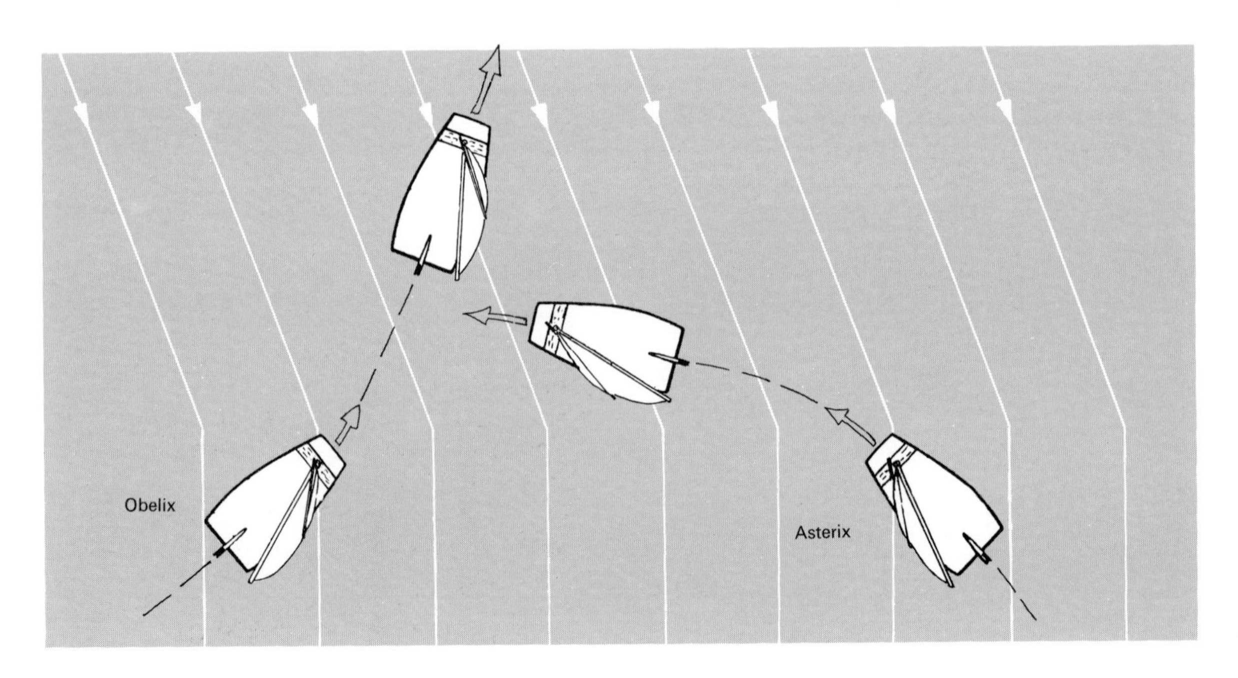

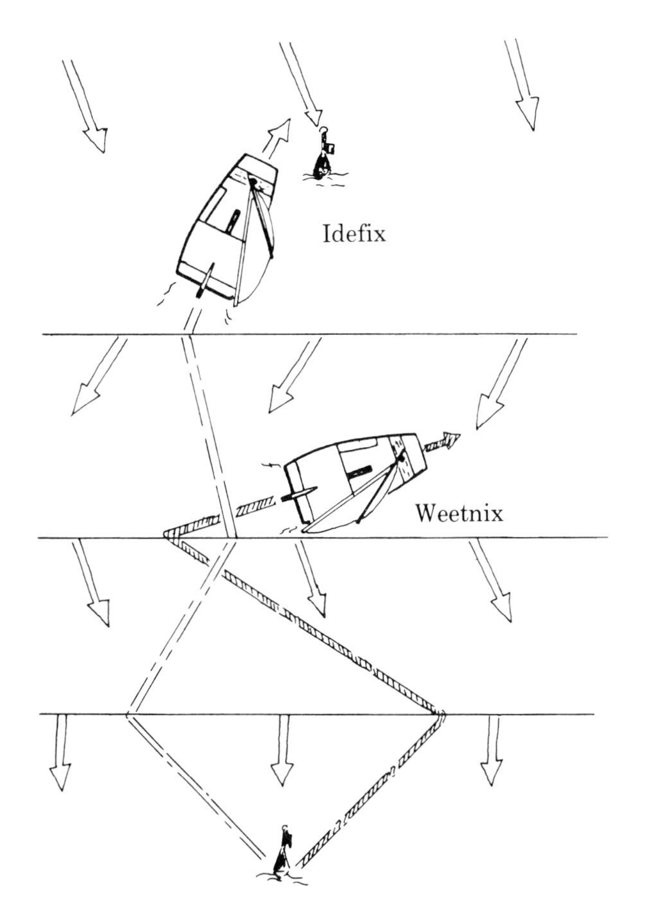

behind **Obelix** although they started together. If the helmsman of **Asterix** is clever he will tack and take advantage of the new wind direction. Generally speaking, you should tack if the wind comes ahead.

In the illustration you can see how much difference it makes if you always take advantage of changes in the wind (wind shifts). Boat **Idefix** always tacks well when there is a change of wind but **Weetnix** always turns in the wrong direction.

The result is clear. **Weetnix** is still only just over halfway when **Idefix** is already at the buoy.

Remember

- **In a gust you can gain speed by bearing away a little and easing the sail.**
- **If you are sailing close-hauled you can also gain distance to windward by luffing up a little.**
- **In a calm bear away, or try to sail around it.**

20
Another way of gybing

There is a way of gybing when running with the wind directly aft, that helps prevent the sail from swinging over too quickly.

Before gybing get down on your knees or squat down. Now steer the boat so that it is slightly by the lee. That means with the wind blowing over the lee quarter (Boat 2 in the illustration). You then gybe the sail over by pulling the whole sheet bundle with the sheet hand. As soon as you feel the sail coming over (Boat 3) steer back again to the new course (Boat 4) and change sheet and tiller hands. Then sit on the other side. (Boat 5).

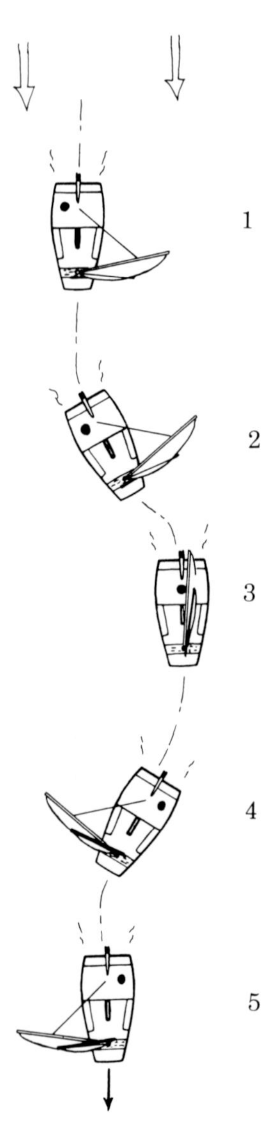

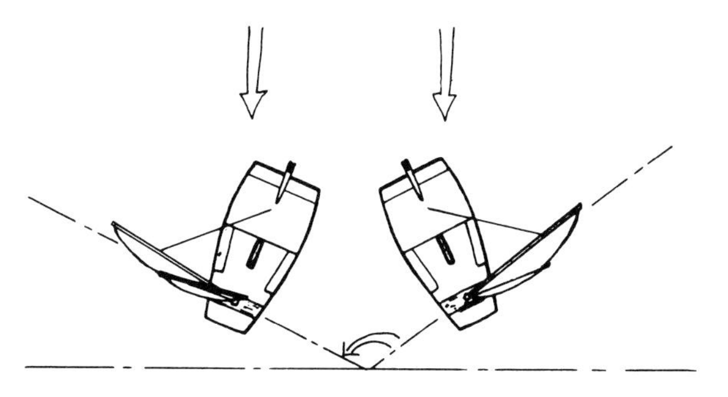

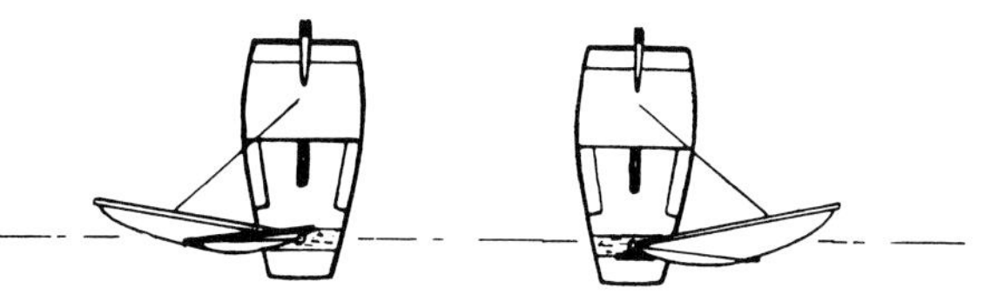

By steering a little by the lee, the arc which the sail must make during the gybe is smaller. As a result the sudden swing is less and the boat more controllable. You can see this in the illustration.

It is best to gybe when you are going fastest. This is often at the end of a gust or on a wave. Also remember to see that the dagger board is almost up during the gybe as if it is blowing hard you will capsize more easily. But be careful not to have it too high as the boom may catch it as the sail comes over.

21
Trimming the boat

Trimming is the art of adjusting everything in the boat to help you go as fast as possible. Not only the sail and the mast, but also your own body is important.

The first section is called weight trimming. That concerns your own weight, the best place to sit in the boat in different wind strengths and points of sailing.

The second section concerns the trim of the rigging and sails.

21.1 Weight trimming

In a boat you can go and sit more forward or more aft or more to windward or more to leeward. You can therefore divide weight trimming into a longitudinal trim and a transverse trim.

First something about the longitudinal trim. If you sit completely aft in the boat you press the stern of the boat into the water. This is called being down by the stern and it has a braking effect, since the boat drags all the water behind the stern. You can see it well if you throw a piece of paper into the water just behind the boat. The paper bubbles along with the boat.

Sitting so far forward that the bow plunges into the water is also bad. The bow then has to push a lot of water away to go forward. As a result the boat goes more slowly. Therefore, the best thing you can do is to sit so that the stern is just level with the water. You can find out where that is if you get someone else to sail along behind you.

When it is blowing hard the boat has a tendency to plunge at the bow. You then go and sit further aft.

Now something about transverse trim. As you were able to read in chapter 7 you should sail the Optimist as upright as possible, that means that the boat does not heel to windward or to leeward. Sailing the boat upright is the first requirement for going fast!!

There are two exceptions to this rule. First, the boat must heel over to leeward a little when there is not much wind. The sail then falls automatically into the right shape. Second, an Optimist is often allowed to heel to windward when sailing before the wind. One reason for this is that the wind is stronger higher above the water and when heeling to windward the sail is higher. The sail will therefore catch more wind. Another reason is that your tiller pulls less and you can hold the boat more comfortably on course.

Remember
- **Stern level with the water.**
- **Almost always sail the boat upright.**

21.2 Sail trimming

The sail is made in a special shape, but you can make it fuller or flatter. You can do this by adjusting the line at the clew, the sprit halyard and the kicking strap. In a hard wind you need a flat sail. You get that by tightening all the three cords.

In gentle winds all the three cords must be looser, but not so loose that creases occur in the sail, running upwards just above the boom.

If you now sit in the right place and have trimmed the sail well, the boat should sail so that the tiller pulls a little. The Optimist is then sailing upright with a little weather helm. A weather helm means that the boat luffs up into the wind of its own accord if you let go of the tiller.

If that is not the case, you must move the mast foot in the mast track. If the boat has a lot of weather helm move the foot of the mast back. As a result of this the sail moves forward on the boat. If the Optimist bears away when you let go of the tiller (the boat has lee helm), move the foot of the mast forward. The sail is then placed further back on the boat.

22
On a lee shore

22.1 Coming alongside a leeward shore

When approaching a lee shore you cannot slow the boat by easing out the sheet, as the wind will be behind you and will continue to fill the sail. Even if you harden in the sheet to reduce the sail area open to the wind you will find that the boat becomes uncontrollable and is moving too fast. But there are ways of drifting into a lee shore with minimum boat speed.

Such a way is to take down the sail and rigging upwind from your desired landing point. You then drift in the direction of the lee shore.

An alternative where you can keep the rig in the boat is to take out the sheet, so that the sail can blow downwind and therefore catches no wind. To achieve this you simply remove the figure of eight knot from the sheet.

When arriving on a lee shore you should remember these things:

1. There may be many breakers on a lee shore. If there are high waves and the wind is strong it may be too unsafe and you should consider an alternative landing point.

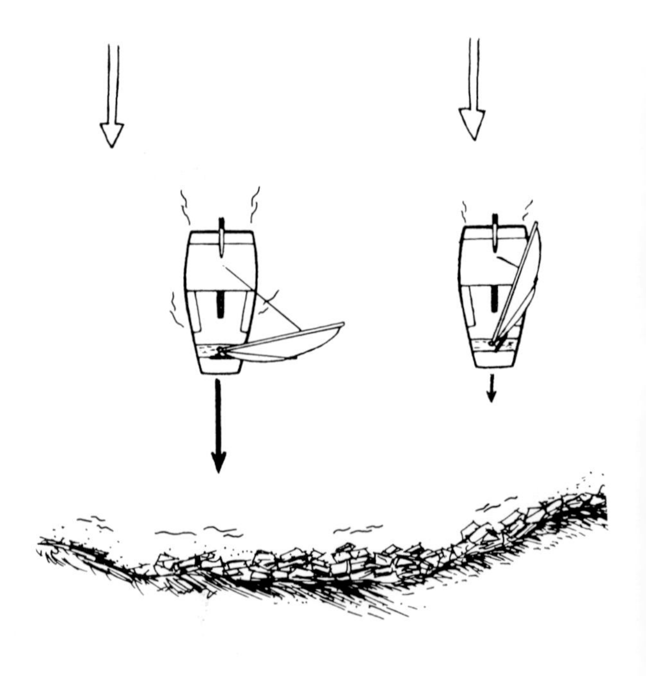

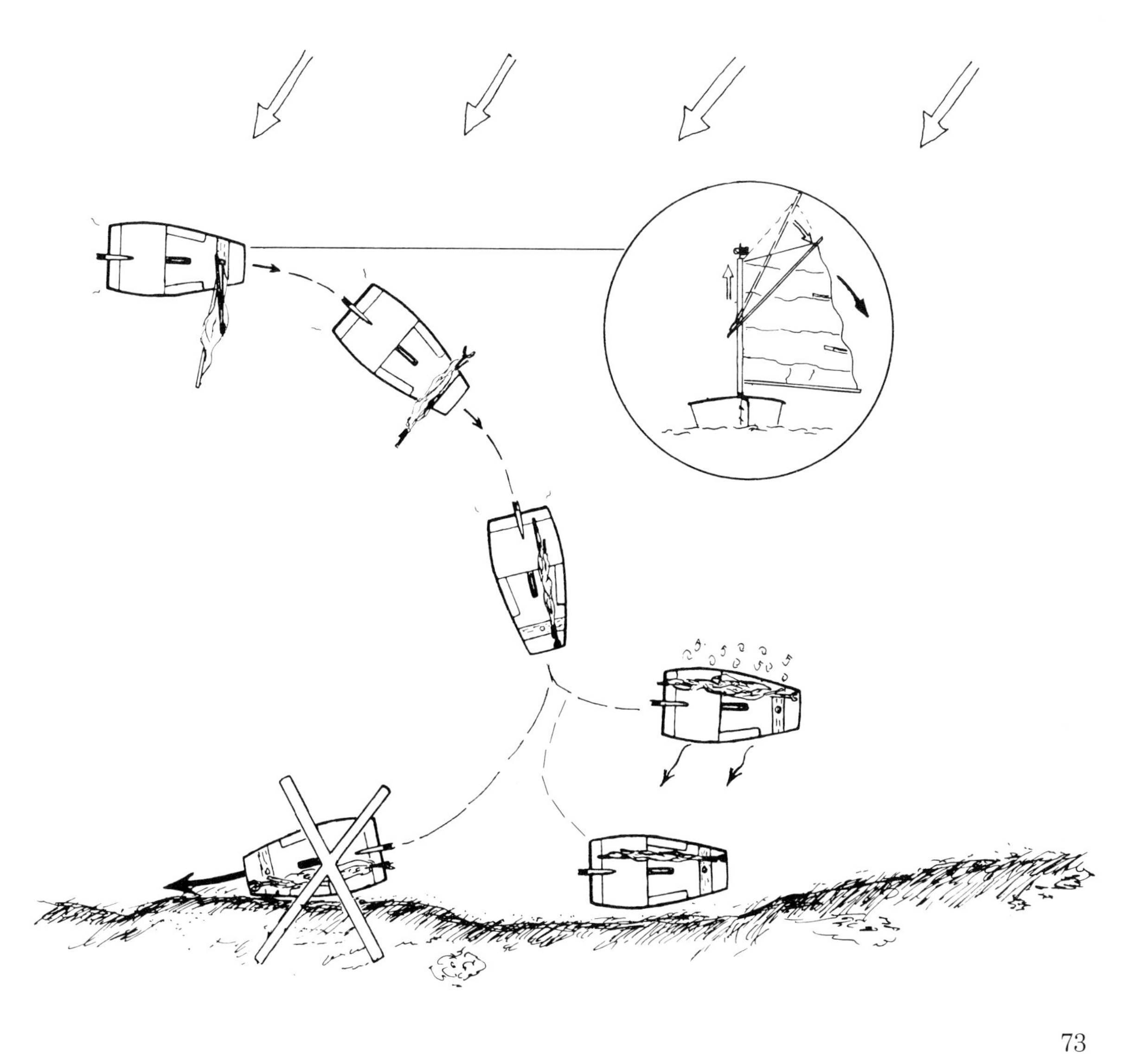

2. As soon as you start drifting towards a lee shore without sail or with the sheet removed, it is impossible to return. You should sail along the shore beforehand to look and see if a landing is possible. For example, you may discover rocks near the shore.
3. When the boat is either beached or laying alongside a lee shore it will be battered constantly by the wind and waves. The boat could be damaged so get it out of the water as soon as you can.

Removing the sail and rig upwind before landing

Heave-to upwind of your landing point with the wind abeam. Untie the figure of eight knot from the sheet and pull it out from the aft block.

The sail will now flap quite freely. The boat stays approximately in the wind abeam position, which allows you to take down the sail and rigging. Pull the kicking-strap off and ease the sprit halyard so that the leech of the sail is no longer taut. This means that the sail catches less wind and the job becomes easier. Lift the rig out of the mast thwart and let it fall with the wind across the boat. Now the rigging is in the boat you can stow it and steer to the lee shore. Remove the rudder and dagger board and get the boat ashore.

When the wind is lighter you could take down the sail more quickly. When the boat is lying with wind abeam ease the halyard and push the boom up (the sheet is out of the blocks and kicking strap loose) and furl the sail around the mast, boom and sprit. Then wrap the sheet round the rigging and secure it with a knot. You can then leave the rig standing upright in the boat and remove it ashore, or take it down and stow it immediately.

Removing the sheet upwind

Heave-to upwind of your landing point. Untie the figure of eight knot from the sheet and pull it out through the blocks. If necessary pull the kicking strap loose and ease off the halyard, so that the leech of the sail catches less wind. Now steer in the direction of the lee shore (see photo overleaf), turn the boat alongside the shore and then quickly remove the sail and rig from the boat. Then take the rudder and dagger board out and lift the boat on to the shore. If you are drifting too quickly towards the lee shore, luff a little earlier and let the boat drift to leeward.

In places where the bank is high this method is impossible, since the boom will hit the bank and the sail will fill. Some piers for example would be too high. Furthermore this is a dangerous approach if it is blowing hard, because the flapping sail will still drive you too fast.

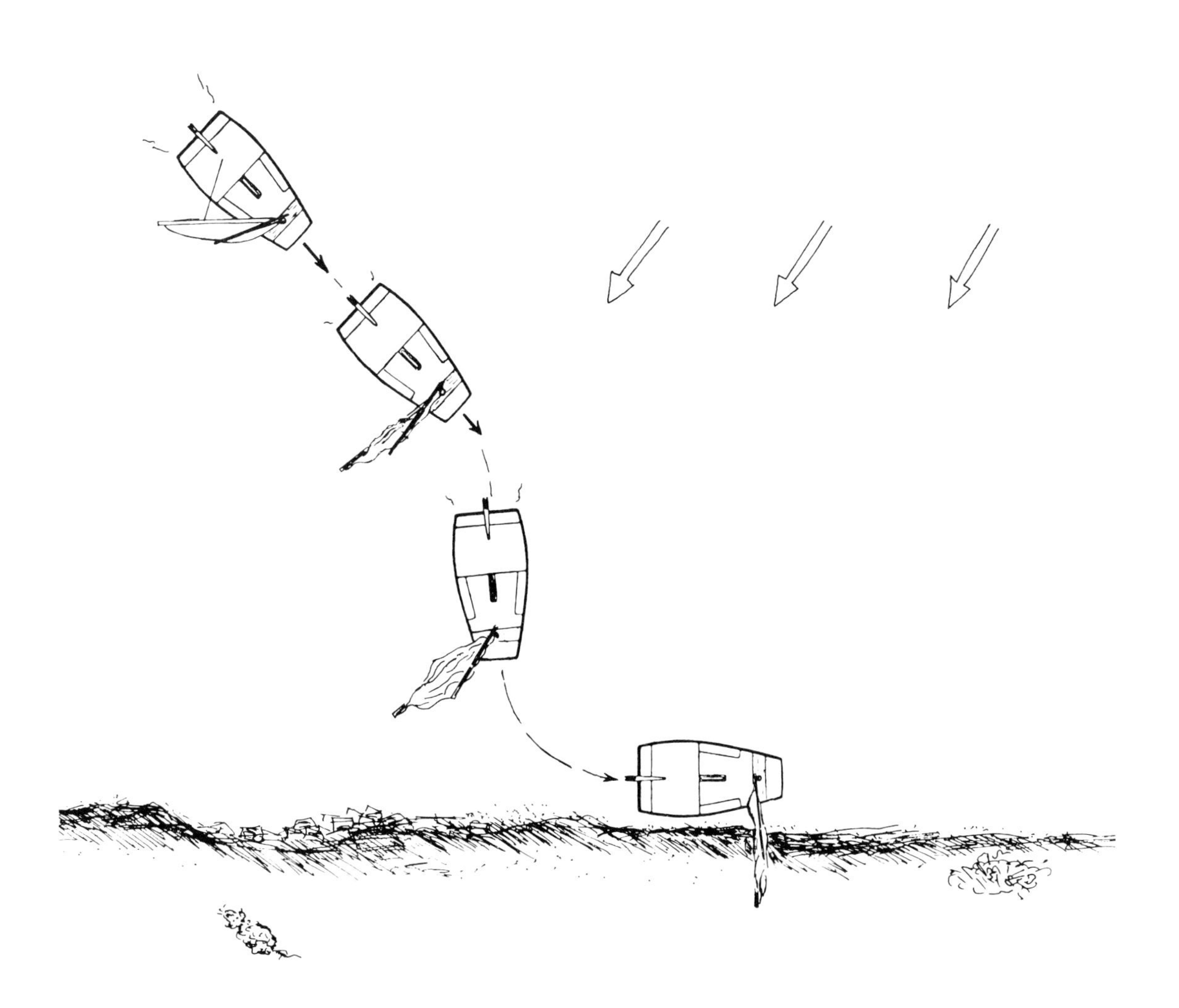

Approaching a lee shore with the sheet removed.

22.2 Casting off from a lee shore

Rig the boat as usual. Be careful to see that the boat lies alongside the shore so that the wind does not come from ahead. As the boom is a good distance out you will probably not be able to fully insert the sheet, so just pull it through the first block and when it is possible through the second block on the boom. The rest can follow on the water.

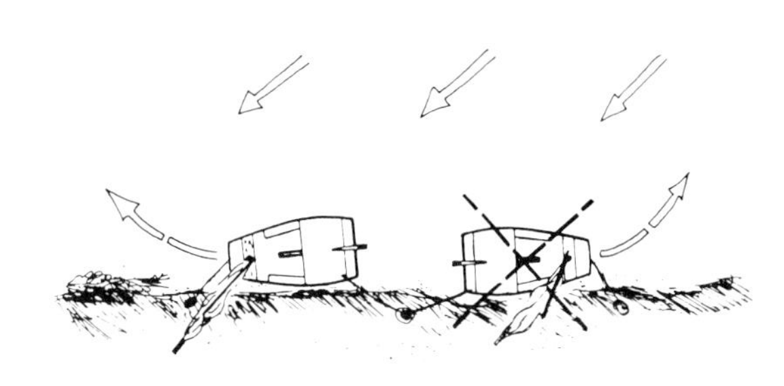

Now it becomes more difficult. You must make sure that the boat is pushed clear of the shore so that you can sail away or the wind will try to push the boat ashore again.

Here are some methods you can use:
1. Go and sit at the tiller and let someone else push you off. In this way the boat is pushed clear and you start with a little speed.
2. If you are alone you can get away by 'stepping'. Stand with one leg in the boat and the other on the shore and push. Make sure that you have the tiller and sheet in hand so that you can sail off immediately. If you don't you will be pushed back to the shore.
3. If it is fine weather you can start in the water. Walk into the water with the boat to a short distance from the shore. Position the boat so that she has the wind on the beam and climb aboard over the stern. You can also use this method of casting off if you are going to sail from a beach, but you must first go deep enough into the water to be able to attach the rudder to the boat.

In the above methods of casting off, you must make good use of the dagger board. It must go as deep as possible, otherwise you drift directly back to the shore again, but do not let it touch the bottom.

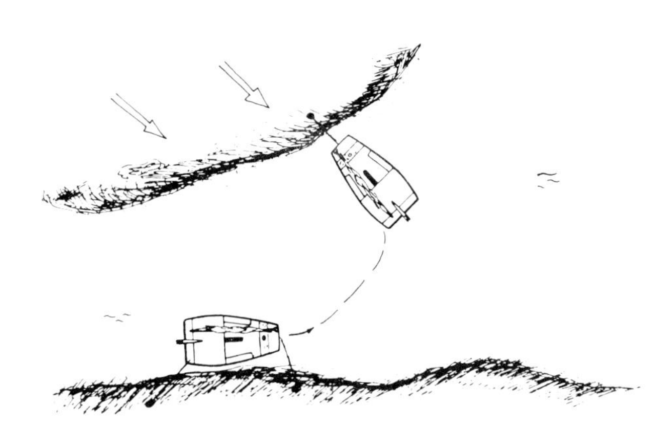

23
The weather

Whenever you go sailing it is important to know what the weather is expected to do. If you are still learning and not entirely confident it is even more vital, since a sudden increase in wind strength can be unpleasant and sometimes dangerous.
Try to listen to weather forecasts and understand basic weather patterns.

These are some useful ways to find a weather forecast:

- The shipping forecasts on the radio.
- Local radio stations broadcast many useful local forecasts.
- The telephone offers many weather services - some especially for yachtsmen.
- The daily newspaper.
- Television news gives a brief, general picture.
- Look at the barometer.
- Look at the clouds and get to know what they indicate.

24 The maintenance of your boat

A difference is always made between maintenance of the boat in the sailing season and in the winter.
In the sailing season you maintain the boat by
- bailing it out;
- cleaning both the inside and the outside with a sponge;
- laying the boat well supported and upside down (then no water or dirt can remain);
- storing the rig and sail dry, with no tension on the sail;
- storing the rudder and dagger board somewhere dry, or putting them away on a flat base.

After the sailing season the boat must have a thorough overhaul. It will then be in good order when the new sailing season arrives. For winter maintenance the following is important:

For a **polyester** hull:
- scrub the boat inside and outside with water and green soap;
- wax the boat with boat wax;
- remove scratches with polyester filler, which you colour over again;
- sand and paint all wooden parts on board;
- remove any tar and oil stains with a cloth soaked in petroleum;
- the boat (preferably in a shed or shelter) should be upside down, supported in several places by drums or spars. This means the boat can get a thorough airing. Also it cannot sag or distort.

For a **wooden** hull:
- clean the boat thoroughly with soap and water;
- sand and paint everything: think especially of the corners and holes where water can gather;
- for storage the same applies as with a polyester boat.

Take the sail off the boom and mast and scrub it clean with soap and water. Next run over the whole sail for loose threads, bad stains and tears. The best thing you can do is to have the sail repaired by the sailmaker, directly after the sailing season.

Also wash the ropework with water and green soap. After that hang it up to dry and store it dry. It is best to replace worn lines at once. Whip or splice frayed ends.

25 Sail racing

Up to this point you have been able to read about how you can sail safely and fast. When, after a lot of practice all these things come naturally you will really enjoy sailing. You can just simply go for a short sail or make real trips without needing to be afraid that you can't get back or that things will happen on the journey that you can't cope with.

However, some people want more. Perhaps you've got the sailing bug so badly that you want to compare your powers of sailing technique with other Optimist sailors. This is possible by taking part in sailing races. In these all boats start at the same time and see who can sail a course laid out round marks the fastest. There are many good books about racing but here is some basic information.

25.1 How can you take part in sailing races?

When races are held, there are usually several classes taking part. The classes start at different times so that boats race against their own sort. Sailing races are organised by water sport and sailing clubs. In order to take part, you must be a member of such a club. You can then take part in the races that your club organises and also in the other races in your neighbourhood. As a beginner it is best to stick to the races organized by your own club. In these so-called members' races, little or no experience in sail racing is necessary. Furthermore, you get to know fellow members with whom you may like to sail more often.

Ask someone who has belonged to the club for some time how to enter the races, you will however always receive supplementary information, such as: venue of the race and the course, starting times, place where you can get race instructions and possibly a course chart and the place where you can obtain still further information. Finally, one more thing. In order to make sure that everyone in a class is sailing with approximately similar boats and sails, there are regulations laid down with which the boats must comply; the class regulations. Thus there are also class regulations for the Optimist. In order to be able to take part in official races, your boat must comply with these regulations. This is possible by having your boat measured by the RYA. If your boat complies with the regulations you will get a measurement certificate.

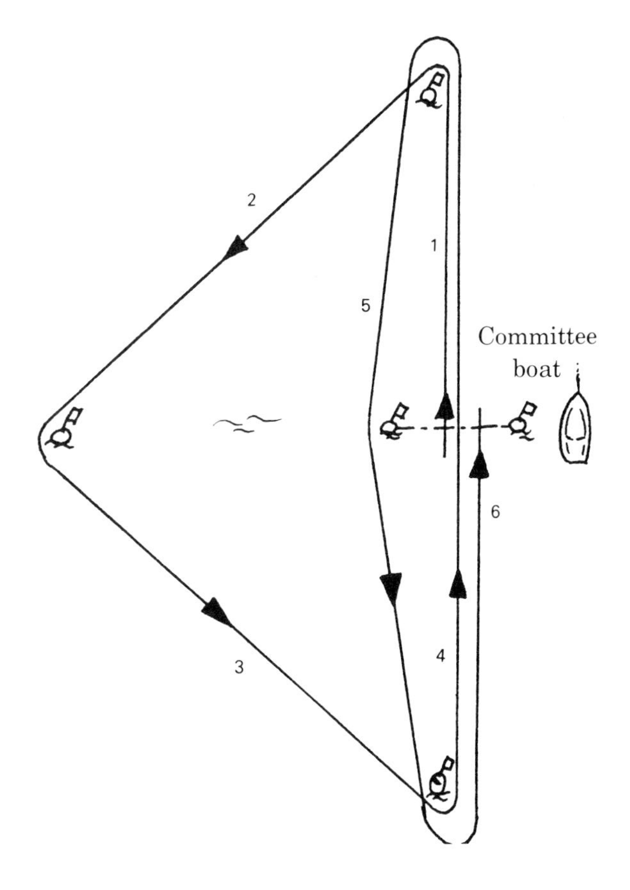

25.2 The course

Races are sailed on a course marked by buoys. You can read in the race instructions which course is to be sailed. Sometimes also a separate course chart is issued.

Most races are sailed on the Olympic course. This is arranged as followed:

You start by sailing over a 'line' which lies between two buoys near the committee boat. This 'line' may also be the finish. The committee boat gives starting and finishing signals. After the start you sail a number of legs from buoy to buoy, namely:

1. Beating leg to the windward buoy.
2. Reaching leg to the gybe buoy.
3. Reaching leg to the downwind buoy.
4. Beating leg, again to the windward buoy.
5. Running leg to the downwind buoy.
6. And finally, another beating leg to the finish.

25.3 Tactics

In racing it is not only a matter of sailing as fast as possible, but also out-smarting your opponents. We call this tactics. In tactics it is a matter of thinking beforehand and thus making the right manoeuvre at the right time. When you are on a beating leg just by another boat, for example, when is it best to tack? Which is the best end of the start line?
A few important rules of sailing tactics are:
- know the rules well (learn the basic racing rules and study the race instructions)
- sail in clear wind
- start well.
All these rules are further discussed in a separate paragraph.

25.3.1 The racing rules

A knowledge of the racing rules is important to decide on tactics. The racing rules are the rules of the game. They are the rules which determine what you may and what you may not try in order to get the better of your opponents. The racing rules also give, just like the rule of the road, rules to prevent collisions.

On the water there is no referee, in fact you are your own and each other's referee. If you commit an infringement of the racing rules then you leave the race of your own accord. If anyone else commits an infringement of the racing rules and you see it, you must protest. That means that you call 'protest' and put up a red flag. After the race you go to the protest committee and make your protest. You then get a form on which you must fill in the details of what has happened. The committee then takes a decision on the matter, going by the racing rules. If the protest is upheld the person against whom the protest was made is usually disqualified.

The racing rules also state how the race must be conducted. For you, as a racing sailor, it is important that you know the different signals for the start and finish of a race.
Alterations and additions to the racing rules for a particular race can be found in the race instructions (for example different starting signals).

Here are a few of the racing rules.

If you want to understand the rules given below, you will have to know the definitions of a few terms.

Going about: a boat goes about from the moment when it is facing into the wind until the moment at which the sail fills again.
Gybing: a boat gybes from the moment when the boom is over the boat until it is completely over the new side.

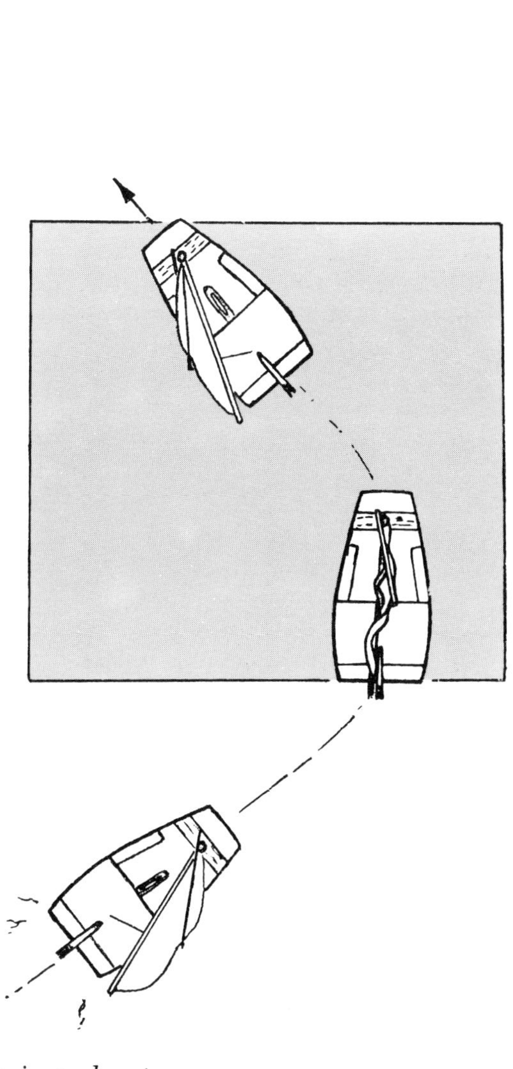

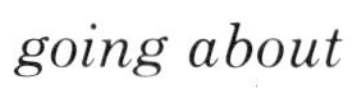

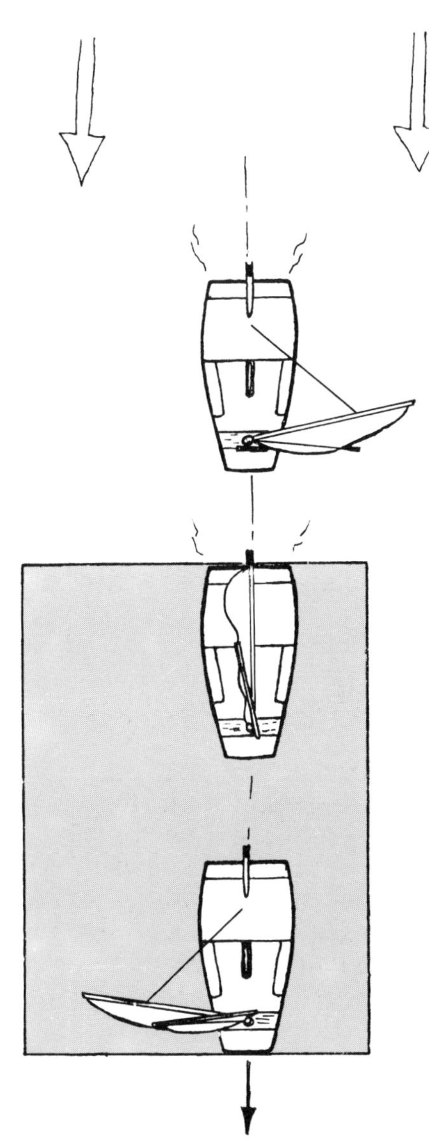

going about

gybing

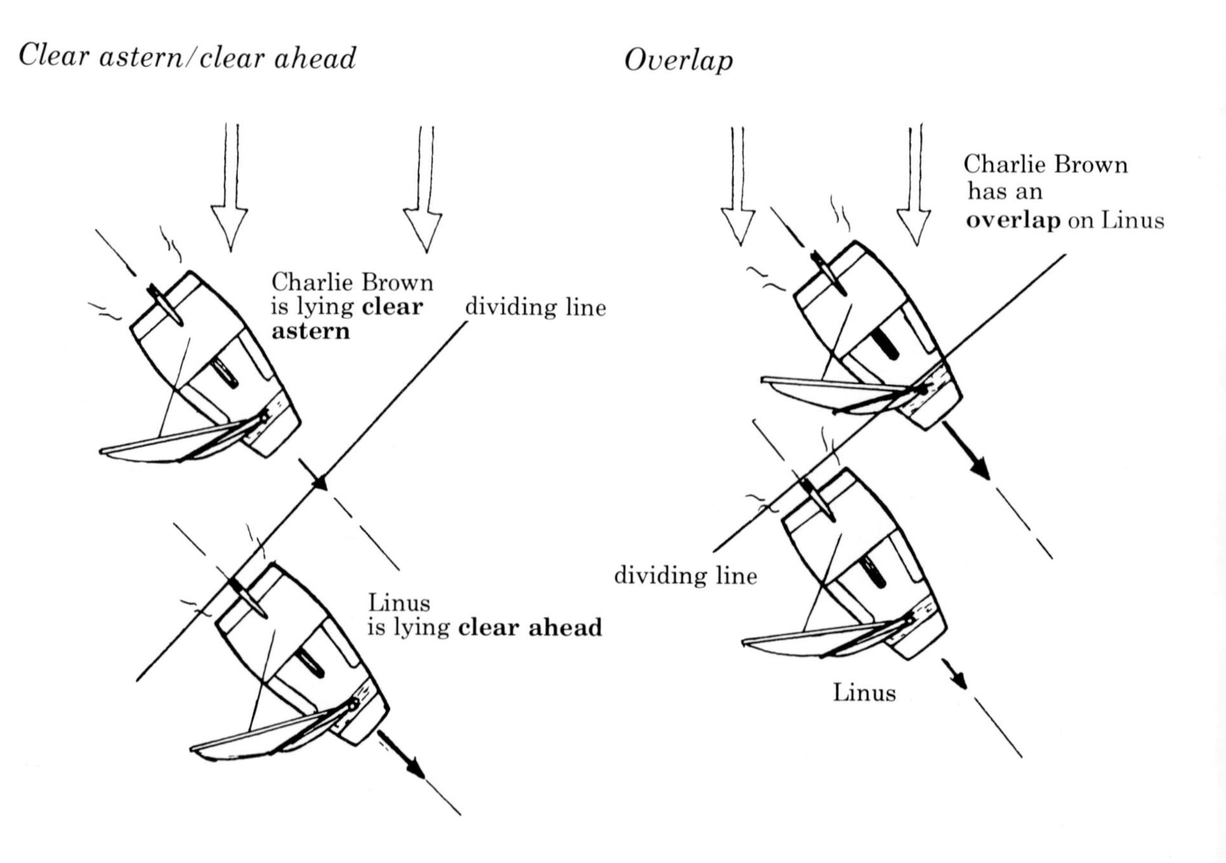
Clear astern/clear ahead
Charlie Brown
is lying **clear astern**
dividing line
Linus
is lying **clear ahead**
Overlap
Charlie Brown
has an
overlap on Linus
dividing line
Linus

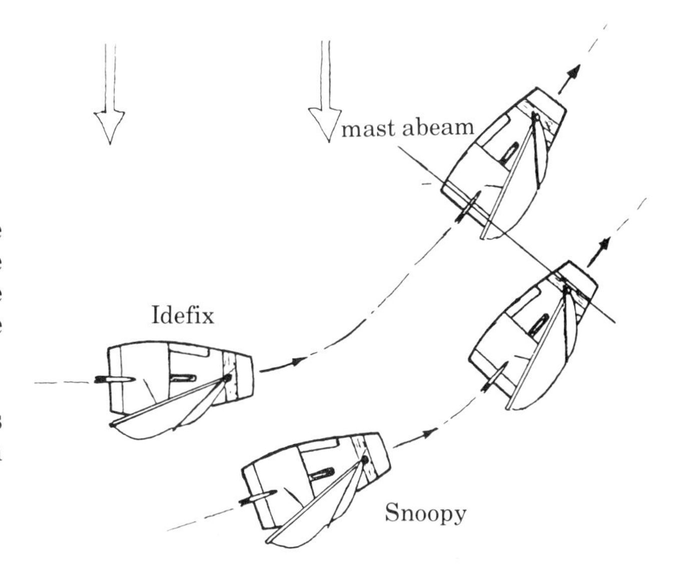

Correct course: the correct course is the course that you would sail, after the starting signal, in order to finish the race as quickly as possible, if you didn't have to contend with other boats.

Here are four interesting situations which may occur during a race. You can read the rules for them below.

1. Converging courses:
- port tack gives way to starboard tack.
- windward gives way to leeward.
- overtaking boat keeps clear.
- the leeward boat may luff up until the helmsman of the windward boat sees the mast of the leeward boat level with him: **Idefix** is trying to pass **Snoopy** to windward. **Snoopy** may luff up in order to counter this. **Idefix** must luff with him till the helmsman of **Idefix** has the mast of **Snoopy** level with him. He then calls: 'mast abeam' and **Snoopy** has to fall back on the correct course. **Snoopy** may only luff up again when the overlap has been broken.

2. Going about and gybing:
- A boat which is going about or gybing must remain clear of a boat which is on a tack. The rules mentioned in 1 only apply when the manoeuvre is completed.

3. Rounding buoys:
- On the same tack the boat has a right to water if it has an overlap inside two boats' lengths from the mark (on the water you must estimate this).

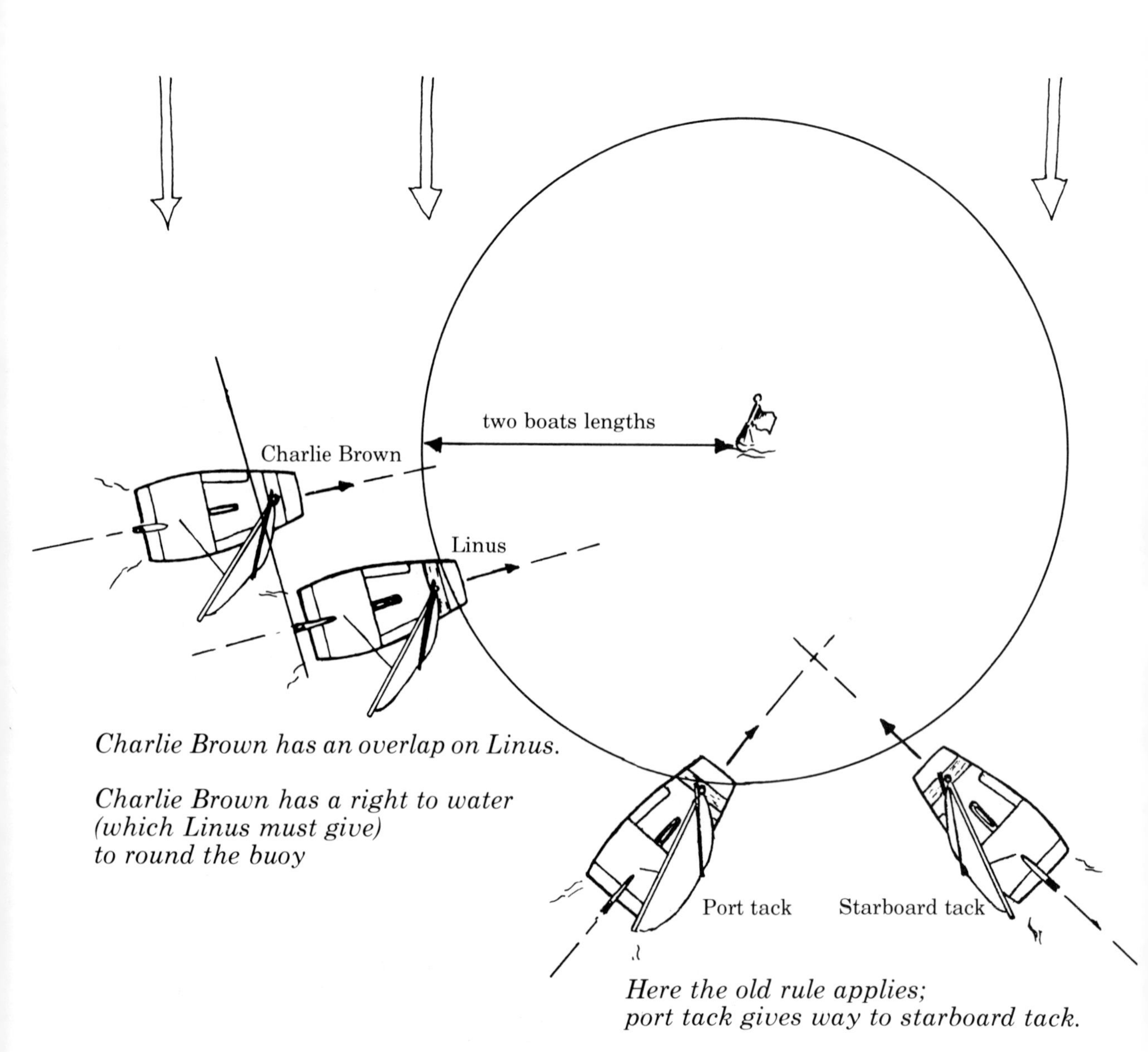

Charlie Brown has an overlap on Linus.

Charlie Brown has a right to water (which Linus must give) to round the buoy

Here the old rule applies; port tack gives way to starboard tack.

- Port tack gives way to starboard tack.

Except in the following situation, at the downwind buoy, where this rule does not apply:

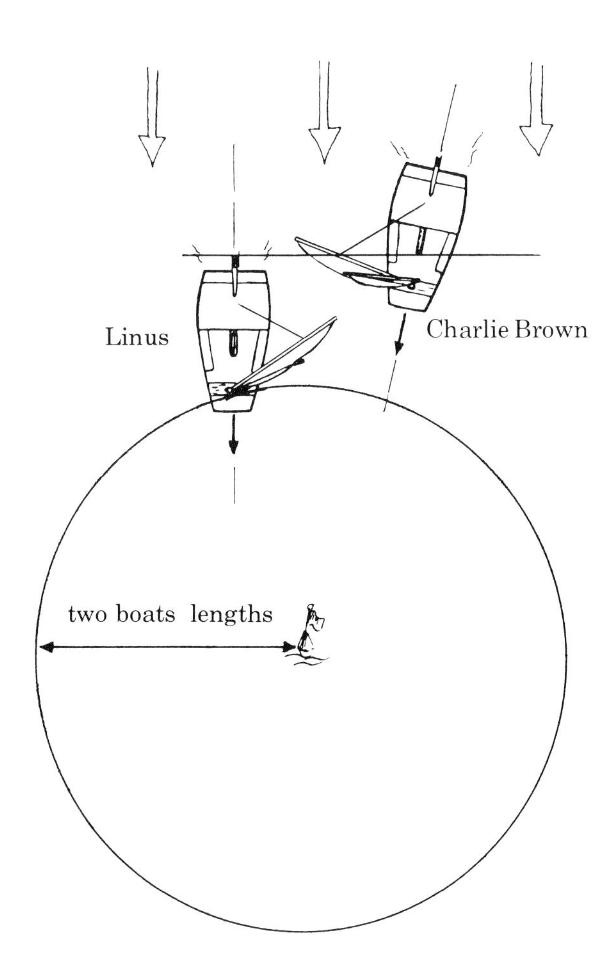

Hitting buoys: If you have hit a buoy, you can redeem the fault by rounding the buoy again in the same direction. You must not hinder anyone in doing so!

4. Starting:
- A boat has started as soon as a part of the boat, the crew or the equipment is over the line.
- The same rule applies to the finish. You have finished as soon as a part of the boat, crew or equipment is over the line.
- If you start too soon you must go back and cross the line again. A recall signal will be shown. You do not need to round the buoys.
- As you are returning to start again you must give way to everyone.

The one minute rule.
- If everyone is called back (general recall) and a one minute rule is imposed you may not cross the start line within the minute before the start - if you do you are disqualified.

Charlie Brown is the inside boat and has a right to water to round the buoy, even though Charlie Brown is on port gybe and Linus on starboard gybe.

- Before the start, the windward rules are different from those described under 2. You may only luff slowly. Also in the mast abeam situation it is permissible to luff, but then no further than to head to wind.

When approaching the starting line, a windward boat at the starting buoy has no right to water at the buoy.

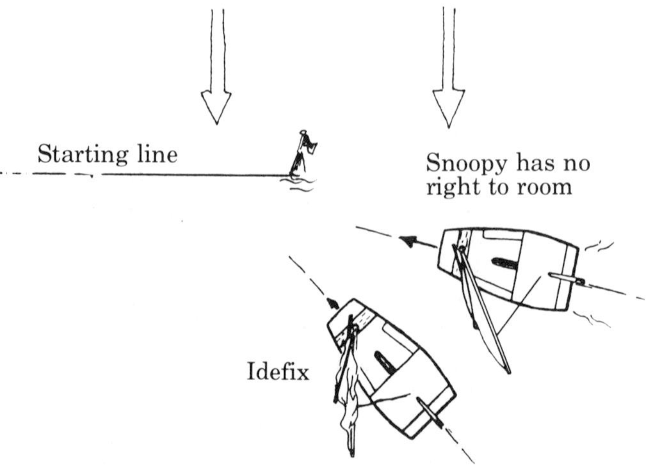

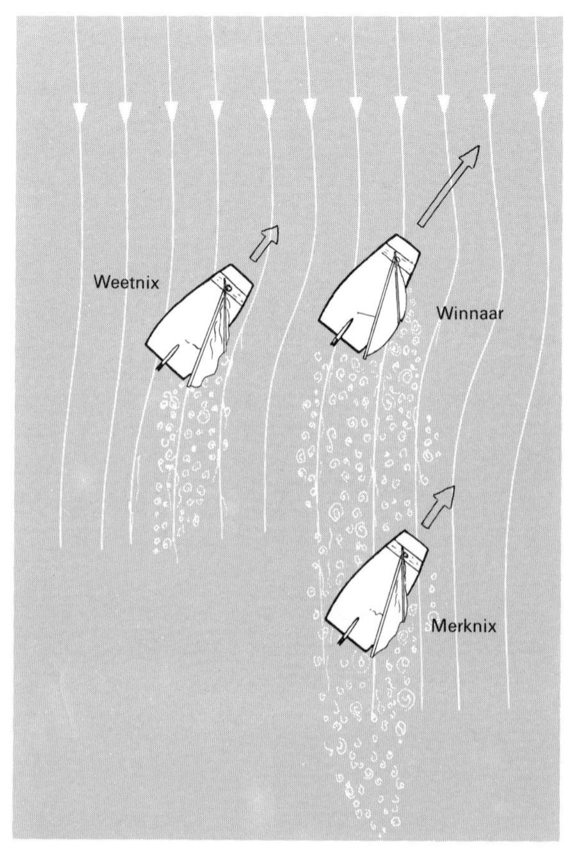

25.3.2 Sailing in free wind

In the lee of trees for example there is less wind, because the trees interfere with the wind. In the same way, the sail of a boat interferes with the wind. In the lee of a mainsail the wind is affected in strength and direction (known as dirty air). The direction of the wind also changes to windward of the sail. What does this mean in a race?

Merknix is sailing in dirty air and will therefore sail more slowly and less close to the wind. There is only one way out: to go about and sail in free (undisturbed) wind. The sail of **Weetnix**

will start to backwind, since the wind is coming ahead. **Weetnix** was already sailing close to the wind so hardening sheet is no help.

Weetnix will also steadily fall further behind and there is only one way out: go about and sail further into free wind. **Winnaar** has meanwhile had free wind the whole time!

Whatever your point of sailing, always try to keep clear of the dirty air created by other boat's sails.

An example where disturbed wind also plays a big role is the rounding of the leeward buoy (see Olympic course). By rounding the leeward buoy well you can sail in free wind and give others dirty (disturbed) wind. **Weetnix** rounds the buoy badly and comes into the dirty wind of **Winnaar.** Furthermore, **Weetnix** sails away from the buoy to leeward of **Winnaar** (see arrow).

You round the buoy well by:

- not sailing close to the buoy, but keeping it level with you at a distance so that you can pass the buoy nearby when sailing close-hauled (**Weetnix** may not come between **Winnaar** and the buoy as she does not have an overlap).
- luff at the moment when you have the buoy level with you, so that you can sail close-hauled alongside the buoy.

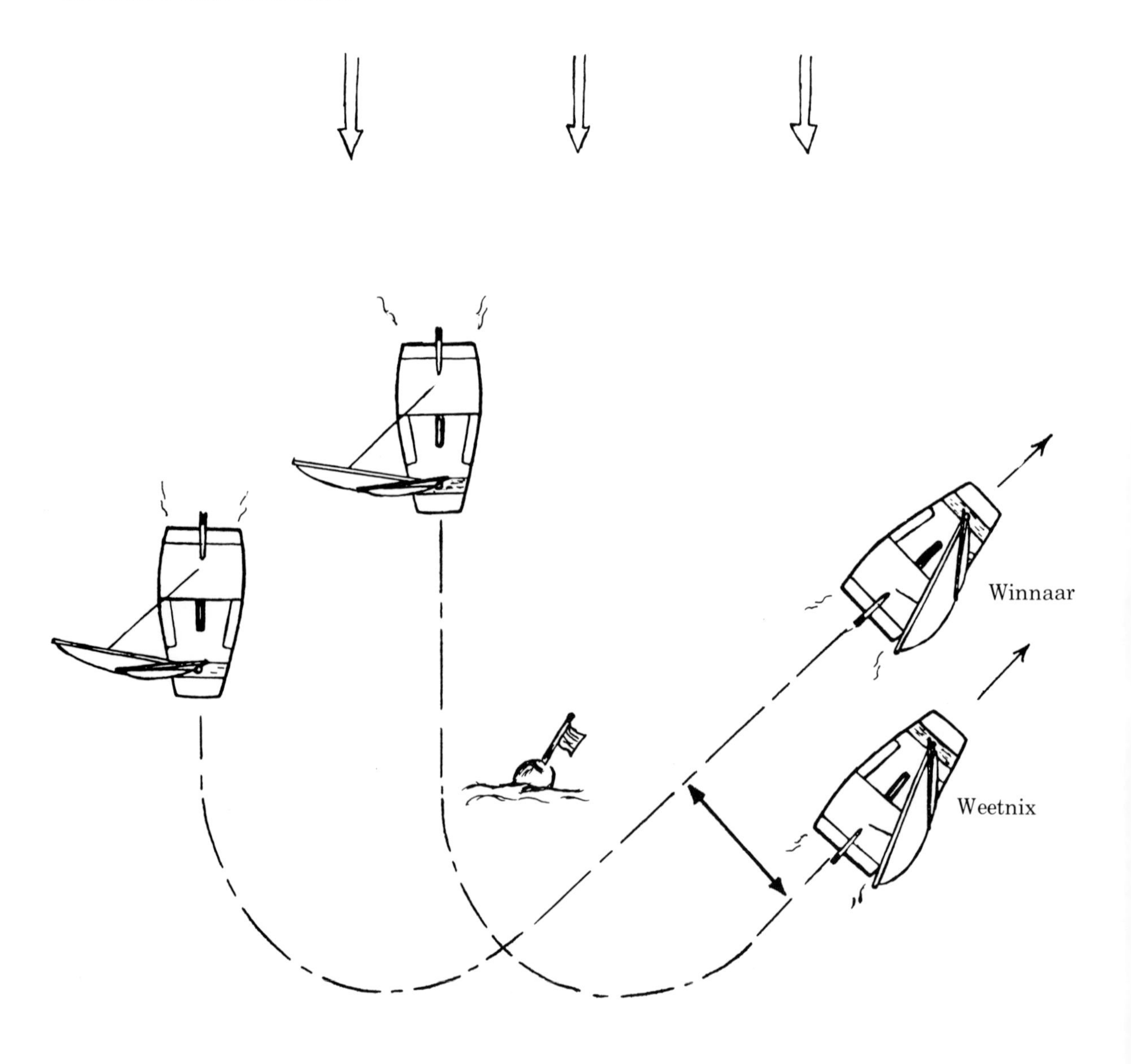
Winnaar
Weetnix

25.3.3. Starting

A good start creates an advantage that will benefit you throughout the race. You will be sailing in clearer air and have fewer boats around you to interfere with your own tactics.

Most starts are to windward (See plan of Olympic course) and there are several things to remember to make sure that you are in the right place at the right time.

Watch out for the sound and flag signals from the committee boat and try

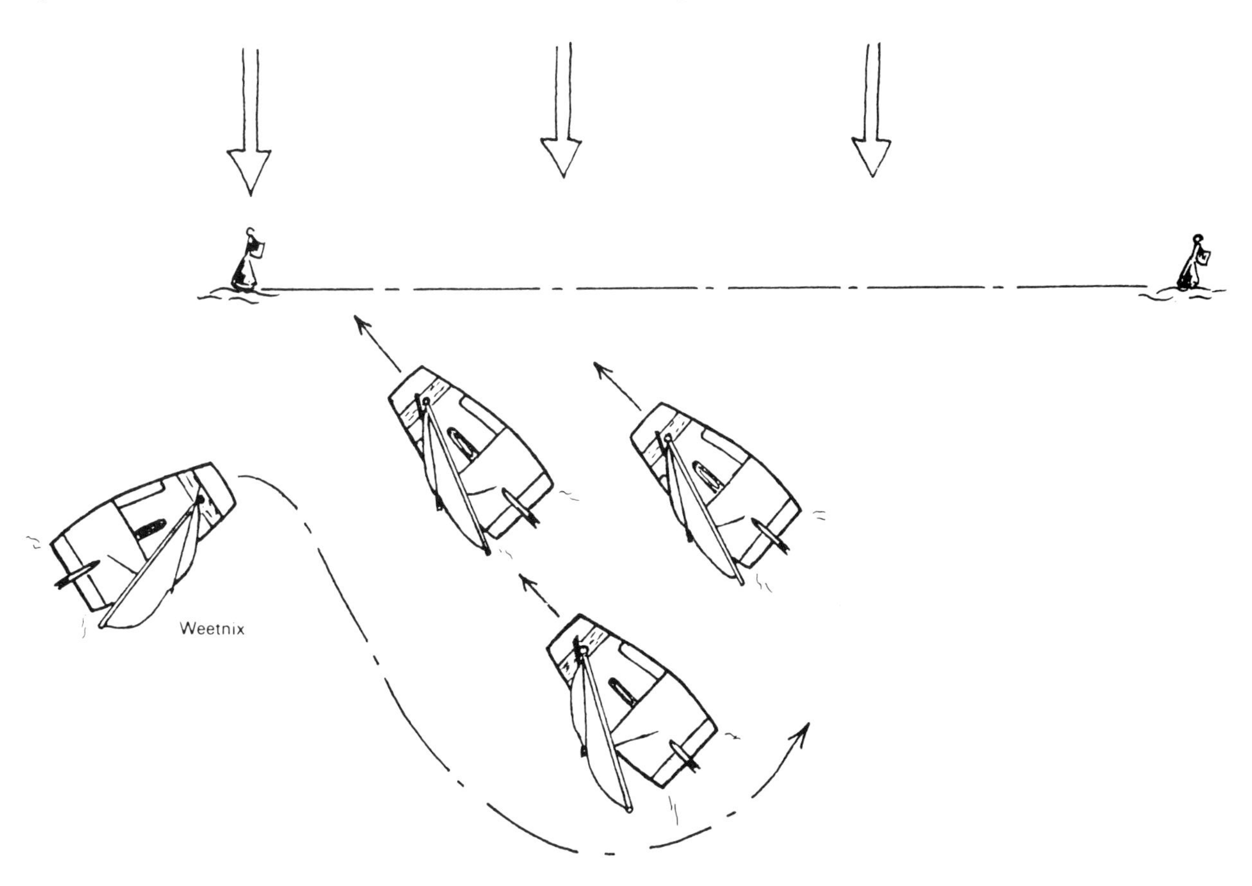

Weetnix is sailing on port tack and must give way to all the boats on starboard.

to keep a countdown of the time to the start so that you can start precisely on the signal.

Cross the line at speed.

Start on starboard tack. It is nearly always faster and safer. See illustration. Start from the point on the line that is closest to the first buoy. (Some tips to help you find that point are listed later.)

Watch out for other boats. Apply the racing rules and try to start in clear wind.

Finding the best end of the line.

Which end of the line is closest to the first buoy? You should find out and use the result to your advantage.

Sail along the line from one end to the other and secure the sheet for maximum performance. Now sail in the opposite direction back along the line with the sail set as before. If the sail starts to luff you will be sailing in the direction of the best end but if you are able to ease your sheet you are sailing away from the nearest end to the first windward mark.

25.4 The roll-tack

If you go about using the technique described earlier in this book you will find that the sail flaps during part of the manoeuvre. During that time the sail is not driving the boat.

There is another method that keeps the sail full for almost the whole tack. This is called the roll-tack.

Both before and after the roll-tack you will be sailing close-hauled and keeping the boat upright.

The roll-tack follows this sequence. (See illustration.)

1. Come inboard a little and ease the tiller; the boat will heel to leeward and luff.
2. When the boat is almost head to wind, lean out again which will make the boat heel to windward. This pulls the sail over to the windward side and keeps it full.
3. At the same moment as the boom comes over, go quickly to the opposite side, transferring tiller and sheet from hand to hand as you do so.
4. Now hang out so that the boat heels over to the new windward side. By doing this you pull the sail over for the second time and therefore the sail is still full.

The whole movement must flow and be quick. You will see that a roll-tack requires a great deal of practice.

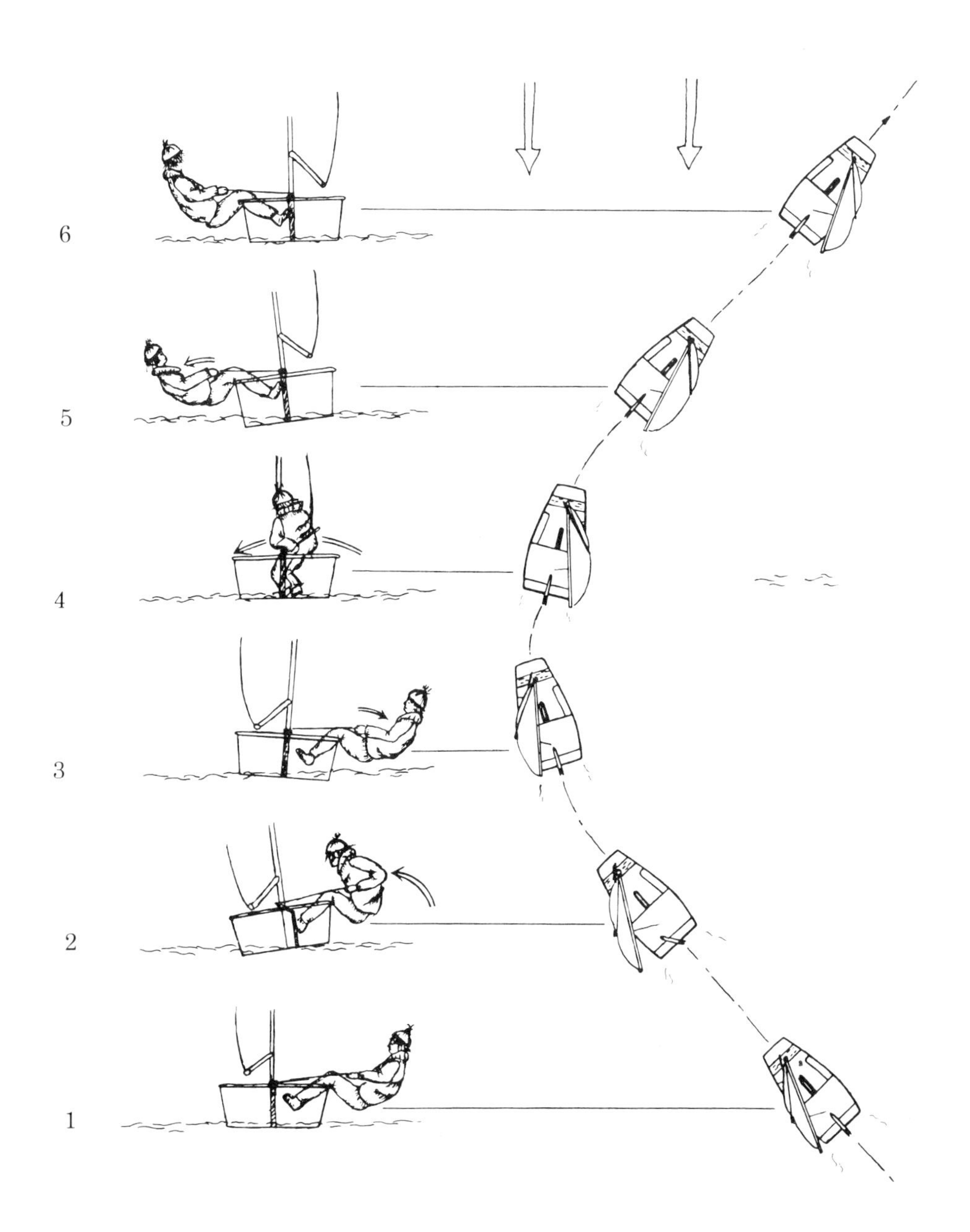
6
5
4
3
2
1